Urban Nature and Childhoods

This book challenges the notion that nature is a city's opposite and addresses the often-overlooked concept of urban nature, and how it relates to children's experiences of environmental education.

The idea of nature-deficit, as well as concerns that children in cities lack for experiences of nature, speaks to the anxieties that underpin urban living and a lack of natural experiences. The contributors to this volume provide insights into a more complex understanding of urban nature and of children's experiences of urban nature. What is learned if nature is not somewhere else but right here, wherever we are? What does it mean for children's environmental learning if nature is a relationship and not an entity? How can such a relational understanding of urban nature and childhood support more sustainable and more inclusive urban living?

In raising challenging questions about childhoods and urban nature, this book will stimulate much needed discussion to provoke new imaginings for researchers in environmental education, childhood studies, and urban studies.

This book was originally published as a special issue of *Environmental Education Research*.

Iris Duhn is an Associate Professor in the Faculty of Education at Monash University, Melbourne, Australia. She has a long-standing interest in critical childhood studies, environmental education, and sociology.

Karen Malone is a Professor of Education and Research Director at Swinburne University of Technology, Melbourne, Australia. She writes extensively about childhoods in the Anthropocene and has published extensively in environmental education research.

Marek Tesar is an Associate Professor in the Faculty of Education and Social Work at the University of Auckland, New Zealand. His published writing focuses on childhood studies and philosophy.

Urban Nature and Childhoods

Edited by
Iris Duhn, Karen Malone and Marek Tesar

Routledge
Taylor & Francis Group
LONDON AND NEW YORK

First published 2020
by Routledge
2 Park Square, Milton Park, Abingdon, Oxon, OX14 4RN

and by Routledge
52 Vanderbilt Avenue, New York, NY 10017

Routledge is an imprint of the Taylor & Francis Group, an informa business

First issued in paperback 2021

British Library Cataloguing-in-Publication Data
A catalogue record for this book is available from the British Library

ISBN 13: 978-0-367-33412-3 (hbk)
ISBN 13: 978-1-03-209125-9 (pbk)

Typeset in Myriad Pro
by codeMantra

Publisher's Note
The publisher accepts responsibility for any inconsistencies that may have arisen during the conversion of this book from journal articles to book chapters, namely the inclusion of journal terminology.

Disclaimer
Every effort has been made to contact copyright holders for their permission to reprint material in this book. The publishers would be grateful to hear from any copyright holder who is not here acknowledged and will undertake to rectify any errors or omissions in future editions of this book.

Contents

Citation Information

The chapters in this book were originally published in *Environmental Education Research*, volume 23, issue 10 (November 2017). When citing this material, please use the original page numbering for each article, as follows:

For any permission-related enquiries please visit:
http://www.tandfonline.com/page/help/permissions

Notes on Contributors

Claudia Díaz-Díaz is a PhD Candidate in the Department of Education Studies at the University of British Columbia, Vancouver, Canada. Her doctoral dissertation focuses on diversity and social responsibility in early childhood education environments through children's relationships with their places.

Iris Duhn is an Associate Professor in the Faculty of Education at Monash University, Melbourne, Australia. She has a long-standing interest in critical childhood studies, environmental education, and sociology.

Sandra Hickey is a Ngunawal woman from the Canberra area who has lived and worked in western Sydney for over twenty years. She has developed innovative research techniques for investigating Aboriginal English using Facebook, and applies this to the teaching of Aboriginal English, culture, and history in her work at Wilmot Public School, Australia.

Riikka Hohti is a Visiting Postdoctoral Research Fellow at Manchester Metropolitan University, UK. Her research is located in the intersections of childhood studies, education, and posthumanist/new materialist theories. Specifically, her research interests address contemporary childhoods, children and digital devices, animals, and zombies.

Riitta-Marja Leinonen is a Lecturer and Postdoctoral Researcher in Cultural Anthropology at the University of Oulu, Finland, where she also received her PhD on the topic of human–horse relationships in Finland. She continues to study human–animal relations within the field of cultural anthropology.

Karen Malone is a Professor of Education and the Research Director at Swinburne University of Technology, Melbourne, Australia. She writes extensively about childhoods in the Anthropocene and has published extensively in environmental education research.

Isabel Menezes is a Professor of Education Sciences at the University of Porto, Portugal, where she coordinates research on citizenship education in formal and non-formal education contexts, and the civic and political participation of children, young people, and adults, especially those at risk of exclusion (based on gender, sexual orientation, disability, or literacy).

John Morgan is a Professor and the Head of the School of Critical Studies in Education at the University of Auckland, New Zealand. A focus of his research is geographical education and how educators can respond to environmental questions.

Fikile Nxumalo is an Assistant Professor in the Department of Curriculum and Instruction at the University of Texas at Austin, USA, where she is also affiliated faculty with the African and African Diaspora Studies Department and the Native American and Indigenous Studies Program. She is a member of the Common World Childhoods Research Collective.

Veronica Pacini-Ketchabaw is a Professor in Early Childhood Education in the Faculty of Education at Western University, Canada. She is a founding member of the Common World Childhoods Research Collective.

Noora Pyyry is a Lecturer and Postdoctoral Researcher at the University of Helsinki, Finland. She researches participation, everyday affectual politics, and 'thinking with' urban spaces in Helsinki and Barcelona. She uses posthuman, non-representational theorization to study the multiple forces that are at work in everyday encounters from which spatial-embodied thinking emerges.

Pauliina Rautio is an Adjunct Professor and a Senior Research Fellow at the Faculty of Education of the University of Oulu, Finland. She utilizes posthumanist theoretical and methodological approaches in studying education and childhoods beyond humanism and notions of development. Her special interest is child–animal relations.

Clementina Rios has been a kindergarten teacher since 2002, with a particular interest in environmental education in its diverse forms. She finished a Master's in Education Sciences at the University of Porto, Portugal, with research that involved children discussing nature and their learning about nature in and out of school.

Margaret Somerville is a Professor of Education and the Director of the Centre for Educational Research in the School of Education at Western Sydney University, Australia. She is interested in creative and alternative methodologies for researching sustainability education from early childhood to schools and community education.

Tuure Tammi is a Postdoctoral Research Fellow at the Faculty of Education of the University of Oulu, Finland. In his PhD, he addressed questions of political education in a school context from an everyday life perspective.

Affrica Taylor is an Associate Professor of Geographies of Childhood and Education at the University of Canberra, Australia. She infuses her geographies of childhood, common world pedagogies, and multispecies ethnographic research with feminist, queer, and decolonizing environmental humanities perspectives.

Marek Tesar is an Associate Professor in the Faculty of Education and Social Work at the University of Auckland, New Zealand. His published writing focuses on childhood studies and philosophy.

Troubling the intersections of urban/nature/childhood in environmental education

Iris Duhn, Karen Malone and Marek Tesar

ABSTRACT
This collection examines why urban environments are key sites for reimagining and reconfiguring human-nature encounters in times and spaces of planetary crisis. Cities constitute powerful and troubling spaces for human-nature intersections. They typically represent the effects of human dominance over nature: humans in control, taming and managing the wildness of 'nature' by domesticating it. Children existing in these mostly adult designed and orchestrated creations are often ignored as city dwellers, along with animals who increasingly migrate into urban areas. Yet cities are also sites of innovation and 'greening', of critical democracy and renewal, with the most innovative cities including those where children co-create urban environments, and where animals and plants are valued as co-city dwellers. As this collection shows, troubling and reimagining these sites for diverse forms and ways of living, including of encounter with the other, and thus what can be learnt and taught through urban nature childhoods, is one possible pathway for working out different modes of being human with the earth.

Introduction

If our species does not survive the ecological crisis, it will probably be due to our failure to imagine and work out new ways to live with the earth, to rework ourselves and our high energy, high consumption, and hyper-instrumental societies adaptively. We will go onwards in a different mode of humanity, or not at all (Val Plumwood 2007, 1)

The idea for this collection on troubling and reimaging the deeply familiar concepts of urban, nature and childhood arose from a study visit to Berlin, Germany, where urban social, cultural, political and ecological initiatives are in constant tension with aggressive capital driven development. In Berlin, like in many of the world's cities, urban innovation (Ferguson 2014) co-exists with gentrification (Walsh 2013), urban food initiatives (Steele 2008), wildlife and cosmopolitics (Hinchliffe et al. 2005; Duhn 2017), brutal politics of expulsion (Sassen 2014), and with childhoods lived in poverty and childhoods lived in affluence (Chaplin, Hill, and John 2014).

Berlin's politics of urbanity are at least partly about the 'renegotiation of the urban commons' (Ferguson 2014, 14; cf. the urban of Detroit, Johannesburg, Hangzhou, Lamu, Curitba and other case study sites of 'urban renewal' in Unesco 2016). Berlin's negotiations have led to discussions and debate, art works, architectural practices, discontent, disobedience and reclaiming as well as a re-imagining of 'the commons' as shared spaces, resources, and shared belongings. In short, Berlin offers a case study of the experience of troubling and re-imagining urban modes of being in a city that is in the midst of remaking itself, and of being remade.

While much is to be applauded about the regeneration of Berlin, to adopt a critical perspective on remaking and being remade involves acknowledgement of its intensifications: of unequal distributions of space, resources and the 'right to belong'. As we show in this collection, children and nature are disproportionately affected by such unequal distributions, as too are the quality and qualities of urban nature experiences. And in terms of environmental education and its research, crucially: what might be taught or learnt there about one's place in the world, and that of others, human and more-than-human.

On a global scale, for example, dwellers, migrants and visitors to affluent cities have seen initiatives that attempt to reintegrate urban nature into public spaces which then become urban nature showcases, such as the New York High Line Park (Millington 2015). Yet overall, urban slums are on the rise. Predictions are that by

> 2030, the cities of the poor countries of the world will house four times as many people as the cities of the well-to-do countries ... the population living in urban slums – the most rapidly growing structure of the urban landscape in the less developed world – will double to almost 2 billion in the next 15 years. The drifting apart of affluent and poor urban environments thus marks the 'uneven globality' of children today. (Schafer 2005, 1027)

This drifting apart happens at global scales as well as within cities. Even within an inner city radius of 10 km, it is highly likely that uneven childhoods in uneven nature co-exist as inner cities become increasingly stratified for various reasons (Smith 2013).

Urban/nature/childhood in this collection: troubling ontologies

What might be done about this? The approach we take in this collection is to start with troubling the intersections of nature/urban/childhood. This is to signal our intention to move beyond sharp categorizations into working within a more porous space where creative intersections of concepts enable enlivened, complex, possibly messy multiplicities of re-imagined urban/nature childhoods. Contributors to the collection engage critically with nature/urban/childhood intersections to explore what constitutes learning with and about nature for children in urban contexts. Collectively they challenge the assumption that cities are places of human and technological dominance over nature, and that childhoods are increasingly lived in human-centred un-natured urban environments.

What motivates us is a concern that the notions of nature and the urban with respect to childhoods are often disconnected parts of the mainstream educational discourse. Indeed, educational scholarship in the field of childhood nature studies traditionally aligns with humanist ontologies, foregrounding phenomenological methodologies where nature becomes a rich resource for children's sensory and embodied learning in the world (Sancar and Severcan 2010; Beery and Jørgensen 2016; Wight et al. 2016). One of our key purposes in this introduction then, is to show how the limitations of a narrow and nostalgic view of 'child, urban, nature' might be overcome, and to reimagine more diverse approaches to education and educational research that extend beyond humanist ontologies. We continue our introduction with a few orientating comments in relation to environmental education, before sketching the contributions to the collection and digging into these issues in the remainder of our editorial comments.

From humanist to post-humanist

In brief, studies with an ontological commitment to humanism in both education in general and environmental education in particular are inclined to reinforce the Rousseau-inspired idea that education is philosophically tied to the purity of nature and the innocence of the child (Baker 2001). The underlying logic implies that effective education in nature happens when young learners are introduced early to nature, preferably nature without visible human habitation, such as a forest (Maynard 2007). It is hoped that if a young child experiences nature, then a future adult who feels deeply connected to the natural world and is less likely to exploit it, should be the result.

Another underlying philosophical idea is that the 'true self' can only be found in contact with nature (Diehm 2002), drawing from Romanticism and the understanding that the natural world is a unified whole (Wulf 2016). In contrast, cities as spaces where nature is seemingly displaced, appear as places of fragmentation. Urban environments are then in danger of producing a version of self that is less 'in touch' with one's true essence due to the lack of holistic nature experiences (Davison 2008). In this paradigm, learning to be truly oneself relies on experiential learning in 'pure' nature. For environmental educationalists and educators, this may mean reconnecting as many children as possible with as much nature in the wild as is still left (Louv 2005). Correspondingly, cities with their highly controlled and unevenly distributed natural spaces, are problematic environments for experiential learning with wild or untouched nature (Dowdell, Gray, and Malone 2011). Yet the fact remains that most children on this planet are growing up in cities (UNICEF 2012). So why do anxieties over a loss of childhood innocence in a globalised, increasingly urbanized world seem to go hand in hand with anxieties over a loss of nature, particularly 'wild' nature (Dobrin and Kidd 2004; Mercogliano 2007; James and James 2008; Bruni et al. 2017)?

Troubling urban/nature/childhood categories is one way of unpicking this question. It requires us to take up the urgent challenge of thinking with concepts that are slippery and over-used – be that 'nature' or 'Nature' – to show the extent to which they have become generic placeholders for anxieties and fears about the future (Heise 2008). Equally, in engaging how to re-imagine urban nature as children's places for learning about self, other and the planet when the humanist educational project itself is reaching some of its limits (Todd 2016), requires critical theory to shed those anxieties about the future that tempt dichotomous responses, such that the educational project becomes imagined and practiced as that of

> making an imagined future safe, of stopping something from happening that looms in the future, of clearing away the present and the past in order to make futures for coming generations. (Haraway 2016, 1)

As we illustrate in this collection, education as an inherently human-centric discipline faces particularly messy struggles ahead, and thus its theorists and practitioners must find new concepts and practices that are sorely needed for species' survival in deeply troubled times (Simms 2009; Bear 2011; Lee 2013; Tesar 2017).

To elaborate, in this collection humans are typically positioned as one species amongst other species on this planet. Humans, animals, plants, microbial life – increasingly life in all its forms – seems 'on the move' to adapt to rapidly changing conditions on earth (Cuomo 2011; Tsing 2012; Haraway 2016). It makes no sense to separate humans/other-than-humans, nature/culture, cities/wilderness when climate change, ocean acidification and spatial fragmentation affect all aspects of every living organism on this planet (Teamey and Mandel 2016).

The unsettling of dichotomies such as culture and nature, or human and non-human, then, continues to create turbulences, intensifications and resistance (Davison and Ridder 2006), because it further destabilises the very ground beneath our feet. If there is no nature outside of culture and vice versa, then how do we educate our children to care for the earth? Or are we already beyond teaching to 'care for the planet' (Heise 2008), and instead should focus on how childhood/nature/urban as concepts need to be transformed, so to afford a better set of grounds for engagement with the changing conditions our species – as one among many – is facing?

Towards a collective response

Through an open call for submissions, we invited scholars to explore such questions, including how education contributes to knowledges about natured childhoods with a focus on their performances in the urban. Scholars from Canada, Europe, New Zealand, and Australia responded, and we've selected for this collection those manuscripts which best help disrupt familiar and naturalised interpretations of our core concepts, as well as expose some of the limitations of well-practiced views on nature that often inform educational debates.

For example, to some, 'green' and 'play' areas have become too easily associated with 'soft nature' and 'innocence' as good and safe places for childhoods in the city. Playgrounds often create illusions of child-friendly urban places and connectedness with nature and natural materials, whereas 'hard' urban spaces are associated with control, danger, surveillance, and with being devoid of nature and thus non child-friendly (Barratt Hacking, Barratt, and Scott 2007). In this collection, contributors responded to such concerns with theoretically informed explorations by pursuing one or more of the following questions:

- What are the intersections of childhood/urban/nature and its tensions, possibilities and risks for educational thought and practice (Morgan, and also Taylor)?
- Which pedagogies become possible when concepts are unsettled, and how do these pedagogies contribute to new conceptual spaces for natured childhoods in cities (Pyyryy)?
- What is urban nature and what may be the potential of 'urban nature' as a pedagogical lens in education for sustainability with children (Rautio, Hohti, Leinonen and Tammi)?
- What kind of theorisations of urban/childhood/nature might be useful in thinking about city spaces as lived in, and enacted as 'places' (Somerville and Hickey)?
- How has the nature/urban/childhood intersection been shaped over time (Diaz)?
- What places for children emerge when urban nature is explored as unexpected, unknowable and as a site of interspecies encounters and cohabitation (Nxumalo and Pacini-Ketchabaw)?
- What do children themselves have to say about 'nature' (Menezes and Rios)?

Opening our collection is a contribution from John Morgan, who examines urban and pedagogical imaginations related to the intersection of childhood/urban/nature. Morgan's focus is on the educational implications of this entanglement, but first he analyses the forces and structures that shape and produce urban spaces. In a nutshell, his argument is that urban pedagogy is linked with the notion of the 'planetary scale' and the vitality of 'life itself'. The processes that shape children's nature in urban spaces are often obscured, including the effects of economic restructuring. By examining both the conceptualisations and limitations of 'urban' and 'pedagogy' in relation to nature and childhoods, Morgan critically questions the hybridity of these concepts and explores the potentiality, and possibilities, of ensuing encounters. To illustrate, given the conditions and horizons of contemporary neoliberal capitalism, we must ask: are urban places a site of struggle, of identifying the relationships with children and nature, searching for an ideal liveable, smart and sustainable city? There is a need, as Morgan concludes, for a much wider discussion about contemporary urban natures and childhoods in education, as well as consideration of diverse possibilities, theories and perspectives, given the lack of attention to global political forces in some posthumanist and 'new materialist' analyses.

With Morgan's argument in mind, our next contribution, from Paulina Rautio, Riikka Hohti, Riitta-Marja Leinonen and Tuure Tammi, asks what does environmental education require, when we recognise connecting 'child' and 'nature' is only ever a partial answer? Working closely with two child-within-nature events, their research positions and analyses these as 'configurations of mutual emergence of matter and discourse, of subjects and objects, animate, inanimate, human, non-/more-than-human'. This allows them to explore, amongst other things, the relation of a child and a seagull, a relation that exposes '"death" as an indication of a society where some "members can be shot and others detained from cursing in the name of becoming human (less animal)"'. Their work exposes and evaluates the connectedness of the two separate units (child and nature) as much as it maps the mutual emergence of children and their surroundings in relation to each other. As the authors argue, by taking as their unit of analysis the 'shitgull' as an event, their approach 'makes it possible to critically review the complex conditions in which similar events arise rather than trying to find fault and inject cure to what are thought of as a priori participants: "child" and "bird"/"nature"'. In conclusion, Rautio et al. offer a child-nature configuration of the 'urban shop' that contests the aforementioned definitions of 'nature': 'urban children can and always do live coexisting with their natural environments in many literally profound ways'.

In our next contribution, Noora Pyyry asks, 'How to examine the complexities of children's meaningful engagement with the city?' She reads children in the urban place and their onto-epistemologies through an intra-active lens, arguing that children play with human and non-human subjects, and objects, alike. Pyyry utilises Bennett's (2010) call for *enchantments*, that is, those events that offer disruptions which open up space for new reflection. In light of this, for children – and other human subjects – being in the city is about making a home for oneself in the world and with the world, as a childhood/urban/nature entanglement would suggest. However, these theoretical and philosophical ponderings lead Pyyry to alternative ways of thinking about pedagogies and different ways of conceptualising learning. In such thinking, definitions, dichotomies and singularities become extremely problematic. For example, Pyyry asks, what *and where and when* are the pedagogical spaces where non-linear and rhizomatic ideas, as imagined in childhood/urban/nature complexities, can exist? Through Ingold's concept of *dwelling with* (which Pyyry uses in order to explore the complexities of childhood/nature entanglements in the urban landscape), she argues for treating urban places as loose spaces, where intra-active play takes place with human subjects, the more-than-human, and things. In conclusion, Pyyry's work underscores the importance of fostering pedagogies that grow rather than diminish children's power and agency, so that actually, their implication in the shaping of urban places is made clear, including through their pedagogical relationships and entanglements with nature as they dwell with/in urban landscapes.

Mindful of the global intersections of various urban/nature/childhood configurations, Isabel Menezes and Clementina Rios examine four Portuguese schools in suburban settings: three sites are close to major cities, and one is situated in a village. Their contribution explores how young children articulate their understandings on what, where and how they learn about nature, to emphasise that young children's participation in group discussions about nature, environmental problems and solutions are necessary if we are to ensure civic-political dimensions of environmental education from children's perspectives. Menezes and Rios argue thåt this focus on children's ability to voice their thoughts on nature, their concerns about environmental issues and their ideas for solutions is largely missing from education and educational research, due to deep-seated assumptions about childhood as a time of innocence. As Menezes and Rios point out, children as young as 5 years of age are capable of speaking about their imaginings of a more caring relationship between humans and 'nature'. The authors suggest that involving children in group discussions highlights the potential of children to become active citizens and to take a central role in re-imagining and renewing a 'common world' that is shared by humans and more-than-humans alike. The authors conclude that transformative pedagogies can only be successful if underlying assumptions about children's capacity to contribute to civic-political dimensions in environmental education are challenged alongside assumptions about human and more-than-human entanglements.

The possibilities and risks involved when animals are introduced to teaching practice are at the centre of the contribution from Fikile Nxumalo and Veronica Pacini-Ketchabaw. Their work analyses the dilemmas of teachers in an early childhood centre in British Columbia, Canada, when confronted with the challenges of managing a burgeoning stick insect population. These forms of urban/nature/childhood encounters bring to the fore the complexities of ethical animal-human relations. The authors describe the focus on the explicit inclusion of pets in the early childhood curriculum to introduce young children to basic biological epistemology. Stick insects have become particularly prominent in many North American programs as insects are perceived to be more manageable than mammals or birds. The authors pose questions that explore the ethical, political and ecological dimensions of child/animal/educator encounters. For example: what to do when children notice that some of the insects are missing parts of their legs? In what ways are child/animal/teacher relations reconfigured when walking stick insect dilemmas are kept visible, in their complexity and their messiness? The authors explore what happens when pedagogies 'stay with the trouble' to engage with the affective and imperfect practices of urban/nature/childhood through ethical, political and ecological lenses.

Moving across continents and towards non-western ontologies and epistemologies, Margaret Somerville and Sandra Hickey's paper invites us to explore Aboriginal and non-Aboriginal perspectives on the emergence of urban/nature/child pedagogies and intersections in a project that sets about 're-claiming' remnant woodlands. Framed by its attention to indigenous issues, the authors attempt to engage critically with a claim by a group of ecologists, that as urbanisation increases globally, indigenous languages and knowledge are being lost in parallel with the loss of species. Their work analyses children's multimodal images and texts, seeking to draw out as much as dwell on alternative storylines and pedagogies that embrace new imaginings for children to potentially name and change their worlds. Located in a school in the highly urbanised area of Western Sydney, with a significant population of Aboriginal and Torres Islanders and children from diverse cultural backgrounds, Somerville and Hickey call for both the transformative and special capacities of entangling environmental education with indigenous ontologies and epistemologies, including the more than human world. Their paper concludes by questioning how children are seeking out wild places, and how in alternative storylines 'urban/nature/child pedagogies […], time and urban space are conceptualised by the children who bring the past into the present, the present into the past, and past/present/future time exist simultaneously'.

Taking intersections in the past as her immediate horizon, Claudia Diaz offers a fresh perspective on children's lives in rural nature in British Columbia from the 1920s to the 1940s. Diaz's extensive research draws on letters from the Elementary Correspondence School, written by children in rural areas to their teachers in the city. Diaz argues that these letters serve to unsettle certain notions of childhood and nature by highlighting that rural and urban were/are constantly reconfigured in children's letters to their school. The letters then, provide evidence of complex interrelations that shape real and imagined senses of place and self. As Diaz argues, urban life as the imagined place of learning had presence and materiality in remote communities, and children's letters demonstrate a strong attachment to far-away places and people even though most of these children never left their rural community. In essence, Diaz's research looks to re-invigorate debates about the 'contours and terrains' of place-based education by challenging a notion of 'place' that is fixed in time and space. Children in rural British Columbia experienced place as *here-ness* and *there-ness*, transgressing the rural-urban divide through the practice of writing letters and engaging in relationships over time with their teachers in cities. For example, many of these children were hard workers who contributed to their family's economies through working the land and being in nature as much as they were learners in and out of schools with a sense of urban life – contributing yet another perspective to our sense of urban/nature/childhood entanglements.

In the final paper in this collection, Africa Taylor argues that many environmental educators have yet to fully engage with the complexities, challenges and most importantly the implications of the Anthropocene. Taylor's view on the interdisciplinary Anthropocene debates is, that they require educators to engage in a paradigm shift in considering what it means to be human, even as the notion of sustainability remains, as Taylor argues, largely disconnected from debates about the Anthropocene. On the one hand, 'well-meaning, stewardship pedagogies do not provide the paradigm shift that is needed to respond to the implications of the Anthropocene'. On the other, and to respond to this concern, Taylor examines possibilities for alternative ways of considering stewardship pedagogies in environmental education, which take into account the consequences and compelling challenges of the Anthropocene. In so doing, Taylor presents an argument for 'common world pedagogies', and how they can move our shared thinking onwards, from not only relational ontologies that recognise inextricable and productive entanglements with other living beings, entities and forces on earth, but also to a worldly response to the Anthropocene based on collective, more-than-human notions of response-ability and agency. For Taylor then, this entangled relation has the potential to nourish a flourishing future for children who will inherit this uncertain future, an argument that we as editors (and educators) see as pertinent to understanding urban/nature/childhood entanglement.

With these brief sketches of each contribution in mind, we now provide a short overview of each node to further frame the authors' discussions, and return us to our opening themes of reimaging and reconfiguring their intersections.

Nature

Nature can often double up for 'other-than-city' while the idea of urban 'wild' nature might appear as an oxymoron (Hinchliffe et al. 2005). At times, in environmental activism, nature as a sociological and ecological concept is taken to refer to wilderness and those untouched (by human civilisation) yet increasingly precious last remnants of 'pure' nature (Pickerill 2008). However, as Pulitzer-prize winning environmental journalist Elizabeth Kolbert (2014) outlines, there is no untouched nature left on a planet that is deeply striated, fragmented and racing towards mass extinction of species. Indeed, the dawning of the Anthropocene, as some argued even before its widespread use of as a term, marks the 'death' (Merchant 1980) or the 'end' of nature (McKibbon 1989; cf. Morton 2010).

Others expand on such qualifications of the vitality of our notion of nature by arguing that it is now an empty signifier devoid of political meaning (Swyngedouw 2011). As such it become increasingly difficult to conceptualise 'nature' in meaningful ways, particularly in relation to economic and political systems that perpetuate global policies and practices of exploitation of natural resources, of people and of animals (Glasson et al. 2006; Heise 2008). Yet it is this insistence on nature's otherness that has driven conservation politics and education, including policies and practices that led to the establishment of protected national parks and World Heritage sites (Zerner 2000). A powerful example of nature-as-otherness in its own right is Bolivia's legislation that gives earth equal standing with humans (Gianolla 2013). As Latour (2014) puts it, 'Nature' is more than one.

In this, certain scholarship has sought to pick up on how understandings of nature are deeply tainted by post-colonial and neoliberal capitalist thinking, which continues to dominate what 'nature' means in post-consumer advanced capitalist societies (Jaffee 2007; Lloro-Bidart 2015). To illustrate, 'Nature', particularly in the urban contexts of well-educated adults with high levels of disposable income, comes to stand for lifestyle choices: when the knowledgeable consumer has the option of purchasing the slightly more expensive Nestlé pod that carries the fair-trade organic sticker. Nestlé, it is argued, continues its well established and largely unchallenged practice of land/people/animal exploitation but diversifies its offerings for consumers with a niche product for the discerning adult. Nature here becomes a differently exploited, possibly more carefully managed resources that allows consumers (current and future generations thereof) to feel that they care. The example drives home the point that nature is an empty signifier at this historical moment for many Westernized adults, who may socialise as much as educate the next generation into such a status quo.

Given this problematique, instead of a search for meaning, it might be strategically more important to pay close attention to the conditions and the practices that create 'natures' in everyday encounters. Such attention generates highly specific mappings of how nature materialises through repetitions, refrains and also through disruptions in places and across time (Grosz 2008), including through education. The historical, and at times deeply romanticised entanglement between childhood and nature (Taylor 2013) adds another layer of complexity to an already unsettled territory. Recently, the 'children in nature' movement, using social media, conferences and publications has had resurgence in public visibility (Bruni et al. 2017).

A key message being promoted here, particularly to early childhood and primary levels of education, relies on an adult sentimentality regarding urban children's loss of connection to nature, the ills of growing up in contemporary society, and the implications for children's lives and their lack of learning about nature. With this popularisation of grand statements about the importance of children's relationship to 'nature', it is timely to consider what influence these views of 'child' and 'nature' might have on the fields of environmental education and its research. Particularly, as these statements are often underpinned quite liberally by a number of key anthropocentric views, for example: '(1) human societies used to be closer to nature, (2) our current way of life is unnatural or distant from nature, and (3) proximity to nature is a question of learning (and teaching)' (Rautio 2013, 449). These sentiments support the perception that humans are not nature and it is possible for some species, namely humans to be more or less nature, connected or disconnected from nature, and superior to or dominant over nature. How is this taught and learned, and more importantly, where and when, directly and indirectly?

Urban

As we show in this collection, cities are central sites for the (possible) reconfiguring of human-nature encounters in the Anthropocene. Cities act as microcosms of a planetary whole, fashioned specifically for human species (Vince 2014). The city as a selection though, constitutes a powerful imaginary, particularly of the human-nature disconnect and therefore brings credence and attention to our seemingly and increasingly 'de-natured' lives. Critiques of city living highlight the effects of the human dominance over 'nature'; humans in control, taming and managing the wildness of the natural world, keeping nature out. Thus cities as urban spaces are a major element of the organisation of space within the Anthropocene, and continuities as much as changes to their configurations serve to illustrate how and why spatial organisation matters for all species.

For UNICEF (2012), a key challenge is that an estimated 60 million people in low income nations leave the countryside every year, such that cities grow globally at around one million new people every week. It is a challenge to comprehend the impact of this for those humans arriving for the first time and those already adrift in the city. As UNICEF reports, unfortunately one-third of all such city dwellers, especially those newly arrived, will start city life on the streets until they can find makeshift housing in slums or transitional communities on the margins or edges of the metropolitan – land that is unstable or leftover, wild or degraded. They often share these spaces with the animals who have also been pushed to the margins of the sprawling urban populous. As an introduction to being on the margins of urban life, it can be exceptionally grim for all, especially children and animals (Malone and Truong 2017).

While urban spaces are densely populated in comparison to rural areas, they can create temporary incidents of increased diversity as species move in response to pressures in their 'natural' environments. But urbanization, by increasing the concentration of humans, is also decreasing biodiversity locally and globally. As cities grow, vital habitat is destroyed or fragmented into patches not big enough to support the complex ecological communities of plants and animals that once lived in those places. Many animals who depend on natural habitats such as forests, fields or wetlands to survive are increasingly at risk, if not already erased from the life of a city. Urban growth has meant that in many cases, more-than-human habitats have been degraded, lost or paved over to make way for the markers of human urban inhabitation: houses, factories and roads. In the city, those plant and animal species that once flourished have become endangered, or extinct, they are literally swallowed up by human habitats. Even worse, those rural areas that have been abandoned are not necessarily left to be places for animals and plants to flourish or the natural ecology to be restored. Instead they are taken over by other interests, typically those of corporations, and depleted even further (Simms 2009).

These pressures create new 'oxymoronic' events, such as the discovery of biodiversity hotspots in inner city environments where highly endangered species appear at a greater rate than elsewhere in the landscape. The emerging patterns are complex and are only beginning to become visible to research. Ives et al. (2016), for example, argue

> Australian cities support substantially more nationally threatened animal and plant species than all other non-urban areas on a unit-area basis. Thirty per cent of threatened species were found to occur in cities. Distribution patterns differed between plants and animals: individual threatened plant species were generally found in fewer cities than threatened animal species, yet plants were more likely to have a greater proportion of their distribution in urban areas than animals. (117)

In this, humans are not the only climate refugees. Life in general is on the move in search of conditions that allow survival (Kolbert 2014). Yet for researchers, especially in high-income nations, the emphasis has often been on child friendliness or healthy cities for children by addressing the quality of the outdoor environment, such as improving recreational spaces, green spaces, young people's alienation, and controlling traffic to make streets safe for young citizens. In contrast, in low-income nations, the focus on child-cities-nature has predominantly been on more immediate issues such as the impacts of poverty, historical and political injustices, climate change and environmental degradation. Our plea? That environmental educators and researchers examine how urban childhoods are lived and located in assemblages of uneasy and messy urban/nature/childhood encounters. While on matters of where to start, our next node and its intersections with those mentioned above suggest a range of areas we summarise in the conclusion to this introduction.

Childhood

Children growing up in urban places and cities are often perceived as the most disadvantaged group in the Anthropocene. Changing childhood landscapes, the retreat of children from playing outdoors, the poor quality of city environments, are all contributing to this shift in what it means to be growing up in urban spaces.

Many of the world's children are growing up in crowded, polluted environments, with limited opportunities to engage with others species due to an increasing degree of factory agriculture, housing and industry. Highly regulated and monitored lives speak to fears and risks of child abductions and abuse. All these elements have contributed to children's capacity to be independent and more freely in urban environments. But these are not just contemporary issues. Abandoning support for a 'natured' childhood, cityscapes have become degraded for decades. Even when asked to reflect on their own childhood, many adults reminisce about having far more freedom in their urban places than children have today (Malone 2016). Yet the argument accorded to this 'new' child-nature disconnect often relies on an assumption that past generations of children had a closer and more intimate relation with the planet, de-emphasising what has been 'a long history of environmental degradation and disconnectedness' (Dickinson 2013, 7) where the experience of being in 'nature' may not have been predominantly positive.

An emphasis on romanticising the lives of previous generations of children normalises what has been likened to a 'perfect' Disneyfied childhood (Taylor 2013), even though significant evidence has revealed that poverty, disadvantage and environmental degradation have had a long lasting and sinister impact over many generations on children's natured lives. Research illustrates that childhood encounters with the 'natural world' are not always restorative, healthy or as spiritually uplifting as nostalgic writings suggest (Malone 2015, 2016). Within an utopian white middle class America, for example, which is most evident in the writing of influential authors and speakers such as Louv (2005), the experiences of children in less developed nations or disadvantaged cities in high income nations are mostly rendered invisible: those that speak of growing up next to high polluting industries, busy highways and degraded landscapes. For these and many children, perceptions of the environment can be those related to fear and uneasiness. For example, Hordyk, Dulde, and Shem (2014, 6) reporting on research with immigrant and refugee children in Canada revealed: 'Nature was not a utopian ideal waiting to be experienced by children' and 'human and animal predators made walks in a forest dangerous past-times' for these children.

These issues are real for children. Children growing up in such urban areas around the world are facing serious danger from pollutants and pathogens in the air, water, soil or food. Street children for instance, who are often invisible in the research on urban environments, can be exposed to a lack of health services, the danger of traffic accidents, child trafficking and abuse. Finding secure and safe places for refuge (to play and encounter nonhuman others) will continue to be of critical importance for them. All of these vulnerabilities are compounded by the detrimental impacts of an acceleration of climate change and other global disasters predicted to increase in the future.

Studies also reveals the quality of urban environments can have detrimental immediate impacts on children's health, and longer impacts. For example, it has the prospect of limiting young people's ecological identities and their sense of connectedness and empathy with the nonhuman world – including the opportunity for kin relations with other worldly folk. How a child engages with spaces therefore differs enormously on the urban environments where one's childhood is located, yet there has been a tendency for universalizing childhood in child-nature research. Educational initiatives to support children and young people to respond to the impending ecological crisis in urban spaces needs to be attentive to the urban landscapes of childhood (Malone and Truong 2017).

Re-worlding the human/more-than-human in environmental education

To summarise, children's encounters in cities are central to how a child learns what it means to be human, including a human who is in relation with the nonhuman world. Urban spaces shape children and children shape the urban; therefore, childhood experiences, and what is and isn't taught and learned through them, should not be separated from questions of the child and other entities with whom they live (Raittila 2012).

The authors in this collection argue that humanist ontologies have largely failed to justice to this complex situation. Diverse narratives of children growing up in cities at this time of the naming of the Anthropocene illustrate that the child's body becomes more than simply a 'naturalized child'; they are a product of the assemblages, associations and relations through which they are connected to the more-than-human in diverse and complex means. By shifting away from the child in urban nature as the only agential body and focusing on the materiality of child bodies as well as the bodies of other nonhuman entities as relational assemblages, the papers in this collection serve to open up new ethical imaginings for children and their encounters with urban environments as natured potentials with future possibilities.

By troubling what constitutes learning through the intersections and entanglements of child/urban/natures this collection calls into question how we commonly come to view and represent the relations of child-nature-bodies in cities; that is, it is not as simple as traditional discourses and practices of environmental education might have us believe. This collection performs the task of unsettling childhood imaginings where children/nature/urban lives exist separately from one another, or are apart from the many other species and living organisms they co-exist with.

Our goal in this collection then, has been to overcome limitations of a narrow and nostalgic view of 'child, urban, nature' and to reimagine other approaches to education through the lenses of each node and their intersections with each other. It seems timely to trouble big concepts such as nature, urban and childhood and to go beyond humanist ontologies that assume humans are exempt from the ecology of the planet. Instead, as we trust this collection shows, we can seek imaginings and practices for different modes of being human with the earth, now.

Disclosure statement

No potential conflict of interest was reported by the authors.

References

Baker, Bernadette. 2001. "(Ap)pointing the Canon Rousseau's *Emile*, Visions of the State, and Education." *Educational Theory* 51 (1): 1–43.

Barratt Hacking, Elisabeth, Robert Barratt, and William Scott. 2007. "Engaging Children: Research Issues around Participation and Environmental Learning." *Environmental Education Research* 13 (4): 529–544.

Bear, Christopher. 2011. "Being Angelica? Exploring Individual Animal Geographies." *Area* 43 (3): 297–304. doi:10.1111/j.1475-4762.2011.01019.x.

Beery, Thomas, and Kari Anne Jørgensen. 2016. "Children in Nature: Sensory Engagement and the Experience of Biodiversity." *Environmental Education Research* Online First: 1–13. doi:10.1080/13504622.2016.1250149.

Bennett, Jane. 2010. *Vibrant Matter: A Political Ecology of Things*. Durham N.C.: Duke University Press.

Bruni, Coral M., Patricia L. Winter, P. Wesley Schultz, Allen M. Omoto, and Jennifer J. Tabanico. 2017. "Getting to Know Nature: Evaluating the Effects of the Get to Know Program on Children's Connectedness with Nature." *Environmental Education Research* 23 (1): 43–62. doi:10.1080/13504622.2015.1074659.

Chaplin, Lan N., Ronald P. Hill, and Deborah Roedder John. 2014. "Poverty and Materialism: A Look at Impoverished versus Affluent Children." *Journal of Public Policy & Marketing* 33 (1): 78–92.

Cuomo, C. 2011. "Climate Change, Vulnerability, and Responsibility." *Hypatia* 26 (4): 690–714. doi:10.1111/j.1527-2001.2011.01220.x.

Davison, Aidan. 2008. "The Trouble with Nature: Ambivalence in the Lives of Urban Australian Environmentalists." *Geoforum* 39 (3): 1284–1295. doi:10.1016/j.geoforum.2007.06.011.

Davison, Aidan, and Ben Ridder. 2006. "Turbulent times for Urban Nature: Conserving and Re-inventing Nature in Australian Cities." *Australian Zoologist* 33 (3): 306–314.

Dickinson, Elizabeth. 2013. "The Misdiagnosis: Rethinking "Nature-deficit Disorder"." *Environmental Communication: A Journal of Nature and Culture.* 7 (3): 315–335.

Diehm, Christian. 2002. "Arne Naess, Val Plumwood, and Deep Ecological Subjectivity." *Ethics & the Environment* 7 (1): 24–38.

Dobrin, Sidney I., and Kenneth B. Kidd, eds. 2004. *Wild Things: Children's Culture and Ecocriticism*. Detroit, MI: Wayne State University Press.

Dowdell, K., Tonia Gray, and Karen Malone. 2011. "Nature and Its Influence on Children's Outdoor Play." *Australian Journal of Outdoor Education* 15 (2): 24–35.

Duhn, Iris. 2017. "Cosmopolitics of Place: Towards Urban Multispecies Living in Precarious Times." In *Reimagining Sustainability in Precarious Times*, edited by K. Malone, S. Truong, and T. Gray, 45–57. Singapore: Springer Nature.

Ferguson, Francesca, ed. 2014. *Make_Shift City. Renegotiating the Urban Commons. Die Neuverhandlung des Urbanen*. Berlin: Jovis Verlag.

Gianolla, Cristiano. 2013. "Human Rights and Nature: Intercultural Perspectives and International Aspirations." *Journal of Human Rights and the Environment* 4 (1): 58–78. doi:10.4337/jhre.2013.01.03.

Glasson, George E., Jeffrey A. Frykholm, Ndalapa A. Mhango, and Absalom D. Phiri. 2006. "Understanding the Earth Systems of Malawi: Ecological Sustainability, Culture, and Place-based Education." *Science Education* 90 (4): 660–680.

Grosz, Elizabeth. 2008. *Chaos, Territory, Art: Deleuze and the Framing of the Earth*. New York: Columbia University Press.

Haraway, Donna. 2016. *Staying with the Trouble. Making Kin in the Chthulucene*. Durham, NC: Duke University Press.

Heise, Ursula K. 2008. *Sense of Place and Sense of Planet*. New York: Oxford University Press.

Hinchliffe, Steve, Matthew B. Kearnes, Monica Degen, and Sarah Whatmore. 2005. "Urban Wild Things: A Cosmopolitical Experiment." *Environment and Planning D: Society and Space* 23 (5): 643–658.

Hordyk, Shawn, Marion Dulde, and Mary Shem. 2014. "When Nature Nurtures Children: Nature as a Containing and Holding Space." *Children's Geographies* 13 (5): 571–588. doi:10.1080/14733285.2014.923814.

Ives, Christopher D., Pia E. Lentini, Caragh G. Threlfall, Karen Ikin, Danielle F. Shanahan, Georgia E. Garrard, Sarah A. Bekessy, et al. 2016. "Cities Are Hotspots for Threatened Species." *Global Ecology and Biogeography* 25 (1): 117–126.

Jaffee, Daniel. 2007. *Brewing Justice: Fair Trade Coffee, Sustainability, and Survival*. Berkeley: University of California Press.

James, Allison, and Adrian James. 2008. *Key Concepts in Childhood Studies*. Los Angeles, CA: Sage.

Kolbert, Elizabeth. 2014. *The Sixth Extinction: An Unnatural History*. London: Bloomsbury.

Latour, Bruno. 2014. "Another Way to Compose the Common World." *HAU: Journal of Ethnographic Theory* 4 (1): 301–307.

Lee, Nick. 2013. *Childhood and Biopolitics. Climate Change, Life Process and Human Futures, Studies in Childhood and Youth*. Basingstoke: Palgrave MacMillan.

Lloro-Bidart, Teresa. 2015. "Neoliberal and Disciplinary Environmentality and 'Sustainable Seafood' Consumption: Storying Environmentally Responsible Action." *Environmental Education Research* 23 (8): 1–18. doi:10.1080/13504622.2015.1105198.

Louv, Richard. 2005. *Last Child in the Woods: Saving Our Children from Nature-deficit Disorder*. Chapel Hill, NC: Algonquin Books.

Malone, Karen. 2015. "Posthumanist Approaches to Theorizing Children's Human–Nature Relations." *Space, Place and Environment, Geographies of Children and Young People* 3: 185–206. doi:10.1007/978-981-4585-90-3_14-1.

Malone, Karen. 2016. "Theorizing a Child–Dog Encounter in the Slums of La Paz Using Post-humanistic Approaches in Order to Disrupt Universalisms in Current 'Child in Nature' Debates." *Children's Geographies.* 14 (4): 390–407.

Malone, Karen, and Son Truong. 2017. "Sustainability, Education and Anthropocentric Precarity." In *Reimagining Sustainability in Precarious Times*, edited by K. Malone, S. Truong, and T. Gray, 3–16. Singapore: Springer Nature.

Maynard, Trisha. 2007. "Forest Schools in Great Britain: An Initial Exploration." *Contemporary Issues in Early Childhood* 8 (4): 320–331. doi:10.2304/ciec.2007.8.4.320.

McKibbon, B. 1989. *The End of Nature*. New York: Anchor Press.

Merchant, C. 1980. *The Death of Nature: Women, Ecology, and the Scientific Revolution*. New York: Harper & Row.

Mercogliano, Chris. 2007. *In Defence of Childhood*. Boston, MA: Beacon Press.

Millington, Nate. 2015. "From Urban Scar to 'Park in the Sky': Terrain Vague, Urban Design, and the Remaking of New York City's High Line Park." *Environment and Planning a* 47 (11): 2324–2338.

Morton, Timothy. 2010. *The Ecological Thought*. Cambridge, MA: Harvard University Press.

Pickerill, Jenny. 2008. "From Wilderness to WildCountry: The Power of Language in Environmental Campaigns in Australia." *Environmental Politics* 17 (1): 95–104. doi:10.1080/09644010701811681.

Plumwood, Valerie. 2007. "A Review of Deborah Bird Rose's Reports from a Wild Country: Ethics of Decolonisation." *Australian Humanities Review* 42: 1–4.

Raittila, Raija. 2012. "With Children in Their Lived Place: Children's Action as Research Data." *International Journal of Early Years Education* 20 (3): 270–279.

Rautio, Pauliina. 2013. "Children Who Carry Stones in Their Pockets: On Autotelic Material Practices in Everyday Life." *Children's Geographies* 11 (4): 394–408.

Sancar, Fahriye Hazer, and Yucel Can Severcan. 2010. "Children's Places: Rural–Urban Comparisons using Participatory Photography in the Bodrum Peninsula, Turkey." *Journal of Urban Design* 15 (3): 293–324. doi:10.1080/13574809.2010.487808.

Sassen, Saskia. 2014. *Expulsions*. Cambridge, MA: Harvard University Press.

Schafer, Wolf. 2005. "The Uneven Globality of Children." *Journal of Social History* 38 (4): 1027–1039.

Simms, Eva-Maria. 2009. "Eating One's Mother." *Environmental Ethics* 31 (3): 263–277.

Smith, Neil. 2013. *Gentrification of the City*. London: Routledge.

Steele, Carolyn. 2008. *Hungry City*. London: Chatto and Windus.

Swyngedouw, Erik. 2011. "Depoliticized Environments: The End of Nature, Climate Change and the Post-political Condition." *Royal Institute of Philosophy Supplement* 69: 253–274.

Taylor, A. 2013. *Reconfiguring the Natures of Childhood*. Milton Park, UK: Routledge.

Teamey, Kelly, and Uli Mandel. 2016. "A World Where All Worlds Cohabit." *The Journal of Environmental Education* 47 (2): 151–162. doi:10.1080/00958964.2015.1099512.

Tesar, Marek. 2017. "Tracing Notions of Sustainability in Urban Childhoods." In *Reimagining Sustainability in Precarious Times*, edited by K. Malone, S. Truong, and T. Gray, 115–127. Singapore: Springer Nature.

Todd, Sharon. 2016. "New Ethical Challenges within Environmental and Sustainability Education: A Response." *Environmental Education Research* 22 (6): 842–844. doi:10.1080/13504622.2016.1164831.

Tsing, Anna. 2012. "Unruly Edges: Mushrooms as Companion Species." *Environmental Humanities* 1 (1): 141–154.

Unesco. 2016. *Culture Urban Future: Global Report on Culture for Sustainable Urban Development*. Paris: Unesco. http://unesdoc.unesco.org/images/0024/002459/245999e.pdf.

UNICEF. 2012. *The State of the World's Children: Children in an Urban World*. New York: United Nations.

Vince, Gaia. 2014. *Adventures in the Anthropocene: A Journey to the Heart of the Planet We Made*. Minneapolis, MN: Milkweed Editions.

Walsh, Shannon. 2013. "'We Won't Move.'" *City* 17 (3): 400–408.

Wight, R. Alan, Heidi Kloos, Catherine V. Maltbie, and Victoria W. Carr. 2016. "Can Playscapes Promote Early Childhood Inquiry towards Environmentally Responsible Behaviors? An Exploratory Study." *Environmental Education Research* 22 (4): 518–537. doi:10.1080/13504622.2015.1015495.

Wulf, Andrea. 2016. *The Invention of Nature: Alexander Von Humboldt's New World*. New York: Vintage Books.

Zerner, Charles, ed. 2000. *People, Plants, and Justice: The Politics of Nature Conservation*. New York: Columbia University Press.

Beyond stewardship: common world pedagogies for the Anthropocene

Affrica Taylor ⓘD

ABSTRACT

Interdisciplinary Anthropocene debates are prompting calls for a paradigm shift in thinking about what it means to be human and about our place and agency in the world. Within environmental education, sustainability remains centre stage and oddly disconnected from these Anthropocene debates. Framed by humanist principles, most sustainability education promotes humans as the primary change agents and environmental stewards. Although well-meaning, stewardship pedagogies do not provide the paradigm shift that is needed to respond to the implications of the Anthropocene. Anthropocene-attuned 'common worlds' pedagogies move beyond the limits of humanist stewardship framings. Based upon a more-than-human relational ontology, common world pedagogies reposition childhood and learning within inextricably entangled life-worlds, and seek to learn from what is already going on in these worlds. This article illustrates how a common worlds approach to learning 'with' nonhuman others rather than 'about' them and 'on their behalf' offers an alternative to stewardship pedagogies.

Since the turn of the twenty-first century, scientists have been warning that the 'Great Acceleration' of human extractive and consumptive activities over the last 50 years has fundamentally changed the earth's geo-biospheric systems, causing the relatively stable Holocene era to tip into the Anthropocene (Crutzen 2002; Steffen et al. 2015). The anthropogenic earth systems changes they identify include: climate change and global warming caused by massive carbon emissions; alterations to the earth's carbon and nitrogen cycles; ocean acidification; and the catastrophic rate of anthropogenic biodiversity loss attributable to urbanisation, industrial agriculture and forestry (Steffen, Crutzen, and McNeill 2007). The cumulative body of scientific evidence of the Anthropocene in turn heralds uncertain ecological futures, profound challenges for the children who will inherit these uncertain futures, and a whole new level of responsibility for educators who are tasked with the job of preparing children to meet these profound challenges. It portends the need for a serious rethinking of the business-as-usual of environmental education, particularly the education of children.

The proclamation of the Anthropocene has spawned a cascade of interdisciplinary debates. Recent years have seen the establishment of a number of dedicated Anthropocene peer reviewed journals and academic book series, and a plethora of peak international Anthropocene-themed conferences. Within the social sciences and humanities, Anthropocene debates have prompted calls for a paradigm shift in thinking about what it means to be human, what we mean by the natural environment, and about our place and agency in the world (Latour 2014; Gibson, Rose, and Fincher 2015; Hamilton 2015). Noting that it is no longer feasible to deny the inextricable enmeshment of human and natural histories, fates and futures, there are mounting calls for a new kind of scholarship and practice that firstly resists modern humanist tendencies to enact the epistemological nature-culture divide that separates our species off from the rest of the world; and secondly to think and act as if we are the only ones that shape the world (Rose et al. 2012; Yusoff 2013; Lorimer 2015).

Given its charter, it is important that the field of environmental education has a discernable voice within these interdisciplinary Anthropocene debates. It is also important for environmental education scholars to respond to the calls of critical Anthropocene scholars to renew efforts for a paradigm shift in understanding human-environmental relations (Chakrabarty 2009; Gibson, Rose, and Fincher 2015). Whilst acknowledging the interdependence of social and environmental issues, and the importance of avoiding anthropocentric attitudes towards the environment, in practice, most environmental pedagogies still pivot around resolutely humanist understandings of agency and position learners as potential environmental stewards.

In the first half of this article, I argue that although well meaning, stewardship pedagogies do not lead us towards fundamentally rethinking our place and agency in the world. To fully engage with the profound implications of the Anthropocene, including the uncertainties of our ecological future, the complex ecological challenges we bequeath to children, and the new onus of responsibility born by environmental educators, we need to move beyond humanist stewardship frameworks and their implicit human exceptionalist assumptions. I detail some of the key feminist critical Anthropocene re-theorisations that reframe our human species as just one of many that make and shape worlds together.

In the second half the article I provide an overview of some of the theoretical and pedagogical work done by members of the Common Worlds Research Collective (2016) that resists the ubiquitous reach of the nature-culture divide – including the divide that positions urban childhoods as antithetical to natural childhoods. This work brings childhood studies, and early childhood environmental education into conversation with the Anthropocene debates within the feminist environmental humanities and more-than-human geographies. Drawing examples from this 'common worlds' research, conducted in urban natureculture environments, I illustrate how common world pedagogies seek to move beyond the limits of humanism and environmental stewardship by acknowledging more-than-human agency, learning with more-than-human world rather than about it, paying attention to the mutual affects of human-nonhuman relations, pursuing more-than-human collective modes of thought, and by learning from what is already happening in the world.

Anthropocene paradoxes and dilemmas

The implications of the Anthropocene are highly contested, as is the name itself. Not everyone agrees about the most appropriate responses. The paradoxes and dilemmas of the Anthropocene are increasingly the focus of analysis and debate (Crist 2013; Haraway 2015; Haraway et al. 2015; Yusoff 2016). Feminists have been quick to point out that one of the greatest dilemmas is that the adoption of this name risks validating human exceptionalism by reifying the 'reign of Man' (Stengers 2013). This, in turn, leads to the ultimate paradox, in which heroic techno-rescue and salvation responses, such as the scramble to find grandiose geo-engineering fixes, simply rehearse the same kinds of triumphalist anthropogenic interventions that disrupted the earth's systems in the first place. Of equal concern are moves that see the naming of the Anthropocene as an overdue recognition of 'Man's' exceptional powers, and as an excuse to redouble efforts to become 'better' at managing the environment, in order to establish what is being referring to as the 'Good Anthropocene' (for instance Ellis, cited in Hamilton 2015). Many regard the 'Good Anthropocene' proposition as the most worrying response of all (Hamilton 2014; Haraway et al. 2015), as it not only acknowledges that humans have exerted destructive power, but it celebrates this power as something intrinsic to being human. It fails to differentiate between human cultures and their radically uneven impacts on the environment. It also fails to acknowledge that not all cultures everywhere see themselves as separate to nature, omnipotent and invincible.

For those of us who have been thoroughly acculturated by humanism's exceptionalist premises, it takes considerable effort to resist the temptation to default back to the comforting belief that we can always find another 'solution' to the problems that we have created. There is a senseless futility in always seeking new techno-fixes or better management regimes. This is because such searches and endeavours perpetuate the circularity of the delusional exceptionalist logic that has created the mess we now face and bequeath to future generations. Something has to change. Along with many other feminist scholars from the environmental humanities and more-than-human geographies, members of the Common Worlds Research Collective approach the Anthropocene naming event as a wake-up call and a moment for intervening in the business-as-usual of everyday thought and action (Instone and Taylor 2015; Somerville and Green 2015; Taylor 2017, Taylor and Pacini-Ketchabaw 2015; LLoro-Bidart 2016;). Such an engagement seizes the proclamation of the Anthropocene as an opportunity to interrupt the suite of 'grand narratives that have led to our blindness' (Stengers 2015a, 12) – whether they rehearse the tropes of human mastery of nature and environmental control and management, or the seemingly benign tropes of human environmental protection and stewardship (Stengers 2015b).

When read as an irrefutable sign of the inseparability of cultural and natural worlds, the figure of Anthropocene can lead us to humility rather grandiosity. When approached as a figure of natureculture entanglement (Haraway 2008) rather than one that confirms human supremacy, it reaffirms the inextricable enmeshment of human and natural worlds, and signals that it is no longer plausible to perpetuate the nature-culture divide that structures western knowledge systems and underpins humanism. This is an important distinction, for it is this same nature-culture divide that leads to the mistaken belief that we can act on nature, at will, and with impunity. Depending upon how we engage with it, the figure of the Anthropocene can interrupt such divisive thinking, and lead to a modest, reflective and ethically attuned response. Changing the entrenched habits of modern western humanist thought, which are so adept at dividing humans off from nature, requires persistence, vigilance and a preparedness to take risks. It is hard work. It requires us to continually interrogate what it means to human, to resituate humans firmly within the environment, and to resituate the environment within the ethical domain (Rose et al. 2012; Gibson, Rose, and Fincher 2015). It also requires us to radically rethink our agency in the world, to understand that we are just one agentic species amongst many, albeit a formidable and potentially destructive one (Latour 2014; Haraway 2015; Tsing 2015), to refocus upon our mutually productive relations with others in this world (Haraway 2008; Alaimo 2010) and to recognise that a precarious and vulnerable environment simultaneously implicates our precarity and vulnerability as a species (Colebrook 2011; Hird 2013).

All of this has specific implications for the field of environmental education. Although Greenwood (2014), LLoro-Bidart (2016), Somerville and Green (2015), Somerville (2017) and Malone, Gray, and Truong (2017) in their introduction to *Reimagining Sustainability Education in Precarious Times* have all called for environmental education scholars to address the Anthropocene, as a field, environmental education has been slow to engage in the interdisciplinary Anthropocene debates and to consider how the Anthropocene's mind-bending complexities, challenges and implications affect its own core-beliefs and approaches. It seems somewhat ironic that sustainability education took centre-stage within the UN Decade of Education for Sustainable Development (2005–2015), for it was during this same time period that scientists were increasingly warning that the patterns of human resource extraction and consumption associated with capitalist-driven 'development' are manifestly unsustainable on a planetary scale and stating that nothing short of a paradigm shift in our thinking about sustainability is needed (Steffen, Crutzen, and McNeill 2007; Steffen et al. 2015). In other words, even though the Anthropocene, as a threshold crossing event, ultimately testifies to the abject failure of sustainable human 'development', and throws up all sorts of pressing epistemological and ontological challenges to modern humanist 'progress' and 'development' discourses, the education for sustainable development policy focus that captured the mainstream of environmental education scholarship during these times, effectively cordoned it off and buffered it from these substantial challenges.

Moves within environmental education

This is not to negate the diversity of approaches on offer within environmental education, nor to deny that some of these approaches, notwithstanding a lack of engagement with the Anthropocene debates, have championed moves to de-centre the human and challenge the nature-culture divide that underpins traditional western separations of human and environmental sphere and issues. Socio-ecological approaches, for instance, have consistently promoted an eco-centric disposition that values the environment for its own sake, in lieu of the entrenched anthropocentric disposition that only values the environment in terms of its usefulness for humans (Wattchow et al. 2014). Intersectionally-attuned scholars and educators who take a critical socio-ecological approach emphasise the interconnections and interdependencies of social, political, economic and ecological systems and concerns (Kyburz-Graber 2013). However, such challenges to the status quo consistently work against the tide of default human-centric concerns and priorities. Articulating reservations about the consequences and effects of the field's shift away from an environmental focus and towards predominantly social sustainability agendas during the UN Decade of Education for Sustainable Development, Kopnina (2012, 699) argues that the privileging of human welfare and economic redistribution within sustainable development frameworks, resulted in 'obscuring environmental concerns' and 'underprivileging ecocentrism'. Other critics claim that because education for sustainability has an essentially human focus, it cannot value the environment for its own sake – within a sustainability framework, the environment can only ever be regarded as a human resource (Hovardas cited in Edwards 2014a, 26).

Anthropocentric and dualistic thinking has also been challenged by environmental educators interested in human-animal relations (Oakley et al 2010). In the editorial for a special issue of *Canadian Environmental Education* called 'Animality and environmental education: Towards an interspecies paradigm', Oakley (2011, 9) condemns 'outmoded' divided frameworks that position animals as 'other' to humans, simply as a way of promoting the superiority of humans. She points out that an understanding of our own animality 'runs deeper than being exclusively 'about' the environment', and not only leads to self-knowledge, but also to 'considering the subjective experiences of other animals' (Ibid, p.9). Others have pointed to the ways in which attending to human-animal relations can shift our understandings of the learning process. From studying how children come to understand their human-ness through living and thinking with other animals, Fawcett (2002) was the first to challenge education's individualistic and human-centric understandings of knowledge production. In lieu of individual learning she proffers the alternative notion of collectively learning within multispecies epistemological communities (Fawcett 2002, 136).

More recently, Malone (2015) has retheorised child-animal street relations in Bolivian urban slums by engaging with Anthropocene-attuned posthumanist philosophies and common world theoretical approaches to the study of childhood. She finds this an effective way of moving beyond the nature-culture binary, but also of thinking about children's environmental learning in ways that do not default to idealised and sentimentalised western notions of children and nature. She is one of the few environmental education scholars who is calling for 'a new imagining of a 'collective ecology' of human and non-human for future sustainability and environmental education' in the Anthropocene (Malone 2015, 20).

Motivated by an engagement with posthumanist theory, Malone's notion of 'collective ecology' is not the same as calls for 'environmental collective action' emanating from political ecology and some of the more activist branches of sustainability education. While the former moves from an understanding of the 'collective' as already constituted by humans and nonhumans alike, the latter assumes that humans need to band together to take collective action on behalf of the environment. As I have already mentioned, education for sustainability has been critiqued for marginalising the environment and re-centring the human through its privileging of social justice agendas (Kopnina 2012). In addition to this, I believe that it also inadvertently reiterates human-exceptionalism through its renewed emphasis upon the transformative powers of collective (and individual) human agency.

The certainty that intentional human agency alone can change the world (for the better) is firmly entrenched within the humanist and activist framings of education for sustainability and/or environmental education, whether or not it claims to be eco-centric or acknowledges the inter-determinacy of social and ecological worlds (Huckle 1996; Stevenson et al. 2013; Wattchow et al. 2014). This is evident in the editors' introduction to the recent collection *Research in Early Childhood Education for Sustainability*, which begins with a quote from Margaret Mead to the effect that 'thoughtful people' are 'the only thing' that can 'change the world' (Mead cited Davis and Elliott 2014, 1) and goes on to identify 'young children and their actual and potential capabilities as change agents for sustainability' as being 'at the heart' of the collection (Davis and Elliott 2014, 2).

The emphasis on young children as potential or actual change agents and/or as future environmental stewards is particularly strong in early childhood environmental education (Blanchard and Buchanan 2011; Chawla and Rivkin 2014; Cutter-Mackenzie et al. 2014; Davis and Elliott 2014; Davis 2015). The notion of the agentic child is heavily promoted in early childhood education. It draws upon the social construction and rights discourses that underpin childhood studies (Prout and James 1990) as well as post-structural interventions that have challenged developmental theory's dominant positioning of early childhood as a passive, vulnerable and incomplete life-stage along the pathway to fully rational and thus agentic adulthood (Dahlberg and Moss 2005). In addition to these influences, the close association of young children with nature can be traced back to Rousseau's figurative 'Nature's Child' legacy and the subsequent Romantic western cultural traditions that perpetuate the view that young children have a special and close affinity with the natural world (see Taylor 2013, 3–57). From this legacy, the assumption is that, if nurtured, children's 'biophilia' (or innate love of nature) will predispose them to become environmental stewards (Wilson 1993, 2011; Chawla 2009).

Such Roussean beliefs still carry traction within the field early childhood education, and implicitly draw upon bifurcations of good nature and bad culture, aligning children with the former. These bifurcations support two assumptions that are pertinent to the urban/child/nature theme of this special issue. Firstly, they assume that nature exists 'out there' in a pure space that is somehow separate to the corrupting cultural/technological/urban domain in which most children grow up. In other words, time spent 'out there' in nature is posited as a necessary antidote to the 'unnaturalness' of children's contemporary urban life. Secondly, and concomitantly, they assume that young children's 'natural' place is in nature, and that the increasing paucity of children's first hand nature experiences in their overly urban lives constitutes a threat to their wellbeing. As Morgan points out in the first article of this special issue, urban geographies that reframe the urban as indivisible naturalcultural spaces offer some fresh insights for studies of children and nature. But there are already early childhood scholars with a particular interest in environmental education who are beginning to critique the default innocent-children-belong-in-good nature position as overly-idealistic and a-political (Elliott and Young 2015) and to contest the assumption that children will automatically connect with nature and learn to care for the environment through free play alone (Cutter-Mackenzie et al. 2014). For the most part, these critiques do not fully tackle the implications of the nature-culture divide, but they do point to the dangers of over-determining what children learn from their outdoor nature experiences and unstructured nature-play. They re-assert the need for environmental educators to play an active role in fostering children's eco-centric and biophilic dispositions, (Edwards 2014a, 25), and in 'engaging children in discussion about the need for sustainability and sustainable actions in their own lives and communities' (Edwards 2014b, 75).[1]

Despite a growing unease with the simplistic positioning of innocent children in good nature as a refuge from bad urban culture, and the emerging internal debates over different approaches, including moves to acknowledge the interdependencies of social and ecological systems and to promote eco-centric sensibilities, when it comes to education's core business of enacting pedagogies, essentially humanist practices prevail. In the main, environmental education reverts to what the teachers (as agentic human students) need do to and the children (also as agentic human subjects) need to learn in order to better care for and protect the environment (simultaneously and ambiguously positioned as holding integral value and as the passive object of human knowledge/needing human care and protection).

As long as environmental education continues to move within the fundamentally humanist parameters of the broader education disciplinary field, I can only see it being restrained by the singular human focus of the dominant social change agendas and continually highjacked by human agency discourses. I cannot see how it will be able to make the paradigmatic shifts called for in the broader Anthropocene debates. Moreover, without directly engaging in the Anthropocene conversations that specifically consider the admixing of urban and natural lives and environments in a 'post-natural' world (Lorimer 2015), and the onto-epistemological implications of the inextricably entanglement of human and environmental fates and futures, environmental education will not have the impetus to systematically interrogate the force field of default stewardship pedagogies. It will not have the resources to support a radical reconsideration of what it means to be a teacher, a learner and indeed to learn in a now-indisputably indivisible world.

Beyond environmental stewardship

At the very time in which the inseparability of cultural and natural worlds is finally, slowly and rudely dawning upon those of us well-schooled in dualistic western knowledge traditions, stewardship pedagogies seem particularly out-dated. Although well meaning, they do not lead us towards radically rethinking ourselves, our place and our agency in the world. Indeed, drawing directly upon a resolutely twentieth century humanist social change agenda, stewardship pedagogies inadvertently rehearse the entrenched sense of human exceptionalism. This is particularly problematic, as it is the mistaken belief in human exceptionalism that has led many in the 'developed' world to think that we can 'improve' upon nature and exploit the earth's resources with impunity. This belief in the radical separation and superiority of 'rational' humans from all other life forms, stems from the modern western epistemological nature/culture divide. It had its roots in Judeo-Christian doctrines, was formalised during the Enlightenment, and institutionalised in the subsequent establishment of the bifurcated natural and social sciences. In short, it is deeply inscribed within the trajectory of the modern western humanist project. Consciously or not, humanist stewardship pedagogies still operate from the premise that humans have exceptional capacities, not only to alter, damage or destroy, but also to manage, protect and save an exteriorized (non-social) environment. The overarching limit to stewardship pedagogies is that they are out of pace with the times. They unwittingly rehearse the division of cultural and natural worlds, not their inseparability.

Beyond education, environmental stewardship stances have historically bought into and perpetuated a number of intersecting and unhelpful bifurcations. Influenced by twentieth century wilderness protection discourses, humanist notions of environmental stewardship reinforce an understanding that pure nature exists 'out there' far beyond human settlement, in an un-peopled and thus pristine state (Cronon 1998). Wilderness stewardship not only suggests that there is there is a 'natural order' to the division between urban cultures and wild natural places that must be maintained, but even more problematically, it renders invisible past-presences. Aboriginal Australian scholars describe wilderness stewardship as a neo-colonial form of 'terra nullius'. They point out that contrary to the binary settler colonial imaginary, the land has never been uncultured, and without its people (Langton 1996). Non-divisive Indigenous onto-epistemologies offer powerful counter-logics to the humanist premises of western stewardship discourses. In the North American context, Indigenous notions of Land have recently been mobilised as a decolonising intervention and a direct challenge to the inherently externalised western notions of place and nature that imbue mainstream environmental education (Tuck, McKenzie, and McCoy 2014). Embedded within the very notion of environmental stewardship is the premise that 'nature' (out there) needs our protection. Ironically, the purer nature is perceived to be, or in other words the more it conforms to the wilderness ideal, the more compelling is the perceived need to protect it from human activities. This, in turn, bifurcates the humans into those 'baddies' who threaten nature and those 'goodies' who seek to protect nature from the baddies.

There is no doubt that intensifying calls for enhanced modes of environmental stewardship constitute the most common responses to the Anthropocene – whether this be via more effective environmental management strategies, the implementation of new forms of environmental damage mitigation, new technological interventions to save nature, or redoubled efforts to cordon off and protect nature, such as those proposed by the 're-wilding' movement (Fraser 2009). However, alongside such responses, there is also a growing chorus of critical Anthropocene scholars who warn against perpetuating human-ism's aporia and conceits by fetishising the phallic figure of modern western Man – the 'Anthropo' of the Anthropocene (see, for instance, Haraway et al. 2015). The central problem with environmental stewardship responses to the Anthropocene is that they risk reinscribing a specifically modern western imaginary of the Anthropo, by operating within humanism's predictable human-centric logics and barely-concealed heroism. It is always ultimately about us – humans to the rescue.

So what lies beyond environmental stewardship pedagogies? How might we reconceptualise our place, agency and learning in an anthropogenically-altered and inextricably entangled natureculture world? How might such reconceptualisations inform new kinds of environmental pedagogies that cir-cumvent the traps of always reverting to the script of humans to the rescue? To begin to answer the first question, I draw upon the scholarship of prominent feminist contributors to the critical Anthropocene debates, whose ideas have informed the common world pedagogies that I review in the next section.

As a prominent feminist science studies scholar since the mid 1980s, Haraway's entangled naturecul-ture reconfigurations and companion species relational philosophies have long provided an alternative to the well-worn routes of humanist environmental activism and humanist environmental pedagogies. Definitely not a fan of humanism and its heroic conceits, Haraway nevertheless vehemently denounces the label of posthumanist. As she puts it, with a characteristically earthy turn of phrase: 'I am a com-post-ist, not a posthuman-ist: we are all compost, not posthuman' (Haraway 2015, 161). Her comments not only pay homage to her disciplinary roots in biology and her interest in the transmutations of organisms, but the *com* of composting is entirely consistent with her long-term commitment to the becoming *with* of lifeworlds. While many posthumanist philosophers have sought to displace human-ism's autonomous and agentic individual subject with the more generative notion of human subjectivity as a continual process of becoming other (most notably, Deleuze and Guattari 1987), it is Haraway who has always insisted that we never do this becoming on our own.

For Haraway (2008, 25), becoming *with* is literally 'a dance of relating' with a whole host of different entities and beings, not all of them human. 'If we appreciate the foolishness of human exceptionalism', she says, 'then we know that becoming is always becoming *with*, in a contact zone where the outcome, where who is in the world, is at stake' (2008, 244). Becoming *with* is therefore an ongoing relational ontological process. Haraway not only insists that we are constantly transforming and being transformed through our real-life, flesh and blood relations with other living beings (not all of them human), she also emphasises that these mutual transformations are also part of a larger process whereby the world itself is transformed. Haraway often refers to these becomings *with* as 'worldlings', as the ways in which the earth's living beings, entities and forces, through their multfarious relatings, make and remake worlds *together*. 'No species acts alone', as Haraway continually reminds us (Haraway 2015, 159). Her relational, multispecies notion of continual world-making underpins common world pedagogies.

Although she insists that we never act alone, Haraway is not denying that the delusional and destruc-tive human-centric impetus that is driving certain actions, particularly those associated with excessive capitalist extraction and consumption, is damaging the earth (Haraway 2015). Nevertheless, she urges that this is not an excuse to revert to heroic human-centric responses. The challenge now, is to learn how to inherit and inhabit damaged worlds by pursuing recuperative responses *in tandem with other species*. Haraway urges us to 'join forces' with other species in order to work towards the collective, and considerably more modest goal of a 'partial … recuperation and recomposition' of our common worlds. It is the possibility of 'renewed flourishing' for all species that she envisions in this collective more-than-human response. As she puts it 'Who and whatever we are, we need to make-with-become-with, compose-with – the earth-bound' (Haraway 2015, 160–161). Haraway's insistence on the 'with' or the 'com' of 'com*panion' species and 're-com*posing' within an already entangled naturescultures world displaces the singular and ultimately binary humanist vision of human agency, caretakership or stewardship on behalf of the environment.

Another feminist science studies scholar, Isabelle Stengers, also emphasises the lively agency of the earth as an assemblage – and one, as she has recently taken pains to note, which is indifferent to our moral dilemmas about it (Stengers 2015a, 2015b). In order to interfere with the conceits of phallic 'Man' within the Anthropocene nomenclature, Stenger's provocatively reclaims the name Gaia to underscore that the earth is a 'multifold creation', not just of 'Man's' making. As a counter-naming strategy, her re-inscription of Gaia (in lieu of Anthropo) 'intrudes' upon an understanding of the earth as "ours' to protect or to exploit' (Stengers 2013, 178). It is important to clarify that Stengers' feminist re-appropriation of the name Gaia has nothing to do with reaffirming the earth as a reassuring nurturing mother who might protect us from looming ecological disaster. Rather, her intention is to remind us (particular those of 'us' who adhere to the phallic 'Anthropo') that that this indifferent earth has 'daunting powers to dislodge 'us' from our commanding position' (Stengers 2013, 178).

In a lecture about cosmopolitics, a concept that expounds the entanglement of science, nature and politics (Stengers 2012), Stengers declares: 'The time is now over when we considered ourselves the only true actors of our history, freely discussing if the world is available for our use or should be protected' (Stengers 2012). She calls for an end to the kinds of human-exceptionalist thinking, politics and practices that set us apart from nature, whether as its masters, its managers, its guardians, or indeed its sole knowledge makers. Appealing to researchers and scholars to risk abandoning established habits of thought, whereby we strive to know *about* the world in order to improve upon it, to better manage it, or to protect it, Stengers proposes that we experiment, instead, with thinking *with* the world. She calls this 'collective thinking' in the presence of non-human others, in order to produce new kinds of 'common accounts' of the world (Stengers 2012).

More-than-human collective thinking is hugely challenging for those of us who are well schooled in human exceptionalism. It requires us to let go of the certainty that humans are the only knowing subjects and the nonhuman world is the object of our knowledge. Thinking with the more-than-human is not something that we can just set our minds to and do. It is a practice that requires a dedicated apprenticeship – a close attunement to what is already going in the world beyond the human – akin to what Latour (2004) describes as 'learning to be affected' by the world. The very process of learning to be affected, to think with and produce common accounts with the more-than-human must, per se, interfere with our human conceits to be the world's only knowing and agentic subjects. Ultimately, this kind of collective thinking and learning also means that those of us who are well-schooled in humanist thinking have a great deal of unlearning to do. It is the multi-layered challenge of simultaneously unlearning old habits of thought, paying renewed attention to how we are affected by the world, and learning new modes of collective thinking that drives common world pedagogies.

Anthropocene-attuned common world pedagogies

The very notion of common worlds is an active, inclusive, more-than-human one, which is borrowed from Latour (2004) but also inspired by Haraway's (2008) generative and collective 'worldings'. More like an aspirational verb than a descriptive noun, *common worlding* or the *commoning of worlds* requires a persistent commitment to reaffirm the inextricable entanglement of social and natural worlds –through experimenting with worldly kinds of pedagogical practice. This means pushing past the disciplinary framing of pedagogy as an a priori exclusively human activity and remaining open to what it might mean to learn collectively with the more-than-human world rather than about it, acknowledging more-than-human agency and paying attention to the mutual affects of human-nonhuman relations.

Bolstered by an engagement with critical Anthropocene debates, and calls from feminist scholars to pursue radical new ways of thinking and living in an anthropogenically damaged world (Gibson, Rose, and Fincher 2015; Tsing 2015), common world scholars make a break with the structuring logics of twentieth century humanist education. They resist the ubiquitous romantic traditions that infuse nature-based pedagogies – particularly in the early years, and the bifurcation of natural and cultural/urban domains – and instead focus upon the hybrid naturalcultural real life worlds that children inherit and inhabit, along with all other life forms. They acknowledge that our messy, mixed up common worlds are imperfect and complex, irrevocably marked by the legacies of anthropogenic environmental damage, and characterized by global inequalities, mobilities, and displacements (Pacini-Ketchabaw and Taylor 2016). They also break with the assumption of childhood innocence, purity, and protection and the concomitant belief that childhood can be cordoned off from the influences of a corrupted and corrupt-ing techo-urban world – typically within an utopian notion of pure and innocent nature (Taylor 2013).

Since the establishment of a common worlds conceptual framework that reconfigures children's lives within the messy, imperfect and inseparable natureculture contemporary worlds that we inherit and inhabit along with a host of nonhuman others (Taylor and Giugni 2012; Taylor 2013), empirical research conducted by members of the Common Worlds Research Collective (2016) has refused the humanist orthodoxies of child-centred learning, and refocussed on studying forms of collective learning that are generated by children's more-than-human everyday encounters, interactions and relations, typically take place in urban natural environments. These include relations between children and place (Duhn 2012; Pacini-Ketchabaw 2013; Nxumalo 2015; Somerville and Green 2015); relations between children and materials (Rautio 2013a, and 2013b; Hodgins 2015; Pacini-Ketchabaw, Kind, and Kocher 2016; Rautio and Jokinen 2016); and relations between children and other species (Taylor, Blaise, and Giugni 2013; Gannon 2015; Malone 2015; Taylor and Pacini-Ketchabaw 2015, 2016).

This growing body of common worlds research increasingly indicates that there is much to be learned about thinking beyond binaries, from studying children's existing more-than-human relations. Outside of formalised pedagogical contexts, close observations of young children's everyday interac-tions with the world around them reveal that many already practice a form of thinking collectively *with* the more-than-human world. For example, when children spontaneously take the persona of other beings or animate and attribute subjectivities to other entities, they are effectively ignoring the onto-epistemological boundaries that divide the world into humans and the rest. This is presumably because they have not yet been fully acculturated into the foundational binary traditions of western education, whereby 'we' (as the superior knowing human subjects) learn how to separate ourselves from the world that we learn 'about' (as the object of our superior knowledge systems).

Counter to the structuring binaries that are embedded within the formalized pedagogical process of humans-learning-about-the-world, children's often-playful encounters with the more-than-human display openness to the 'becoming *with*' dance of relating of which Haraway (2008) speaks. Regardless of the ways in which our human-centric educational frameworks position children – as individuating human beings developing rational autonomy, as individual learners about the world, and/or as indi-vidual change agents in the making – children's actual worldly relations far exceed our binary schema. From their boundary-blurring and category-confounding worldly relations, it appears that children's subjectivities are not solely constituted by divisive and individuating schema. At least part of the time, children appear to be immersed in the process of more-than-human collective and relational onto-logical emergence.

By observing what is already happening when children interact with the world around them, outside of formal educational contexts, Pauliina Rautio's research highlights that we have a lot to learn from children's existing ontological relations. Her Finnish studies of children's ordinary everyday more-than-human encounters offer a window into a world of relational learning that cannot be conceived within conventional humanist developmental frameworks (Rautio 2013a, 2013b; Rautio and Jokinen 2016). From her finely attuned observations, Rautio notes that the things that matter to children are not always the same as the things that matter to adults. For example, by paying attention to the stones that many children carry in their pockets (a common phenomenon that attracts no pedagogical consideration within conventional educational frameworks), Rautio intimately details how children's worldly relations are constituted in inter- and intra-action with the very materials that matter to them (Rautio 2013a). Always hopeful that children's existing worldly relations might provide a non-dualistic experience of the world, and buffer them from the constraining influences of humanism, she describes the moment at which children jump into a snow-pile as fleetingly offering them the clustering, collective identity of 'snowpilechildren', and speculates that this collective ontological experience might momentarily free them from only ever being known as individual human subjects, identifiable by their developmen-tal phase (Rautio and Jokinen 2016). Through her close observations of children's everyday relations, Rautio's research repeatedly shows that children's intimate, immediate and embodied impulses to touch and become with others in their more-than-human common worlds is nothing like the rational quest to know about the world from a distance that characterises western epistemologies. Her meticulous descriptions of children's significant everyday interactions with the nonhuman world articulate a very different kind of embodied and relational learning.

My multispecies ethnographic studies[2] of children's encounters with wildlife in an Australian urban bushland setting reveal how entangled interspecies relations in the 'contact zone' (Haraway 2008, 244) can be generative of the kinds of more-than-human collective subjectivities that remain off the radar of the individualistic and human-centric developmental imaginary. One of the most significant interspecies relationships I observed during this research, unfolded over a 10-month period of weekly walks on a Canberra university campus. The relationship was between a group of pre-school children and a mob of resident kangaroos. Both the children and the kangaroos exhibited a high degree of mutually curiosity. On every walk they spent a considerable period of time studying each other intently. As the 'dance of relating' (Haraway 2008, 244) of these weekly face-to-face encounters became more and more routine and predictable, the children developed enough confidence to move closer to the kangaroos, and the characteristically shy kangaroos developed enough habituated trust not to turn and hop away (Pacini-Ketchabaw and Taylor 2015).

Clearly stimulated by their increasing familiarity and affection for the kangaroos and their close-up observations of these wild animals' embodied modes of being, the children were increasingly curious about what it would be like to live in a kangaroo's body, to listen attentively with large swiveling ears, to be tucked up like a joey in a furry pouch, to rest upright upon an enormous tail. They frequently expressed a sense of kinship with the joeys. On a regular basis the children spontaneously became kangaroos, simulating the kangaroo mannerisms and movements that they had observed so many times in their up-close, face-to-face meetings (Taylor & Pacini-Ketchabaw 2016a). They were, in effect, performing the kind of learning *with* that proceeds from the unfolding of real-life, inter-subjective, inter-species ontological relations, and which is all about actively seeking the kinds of cross-species identifications and inter-subjective 'becomings with' that the divisive humanist learning project, with its structuring subject-object knowledge relations, cannot envision.

It can be tempting to think about children's relations with other species, such as these between the campus children and kangaroos, as only ever charming and innocent. In reality, however, this is rarely the case. Living in the interspecies contact zone can be difficult and challenging for all involved. Multispecies cohabitations are increasingly characterised by intensifying pressures, particularly in those places in which migrating species are thrown together by the forces of anthropogenic environmental degradation, urbanization, deforestation, climate change and biodiversity loss. The back-story of the campus-residing kangaroos in my research is a case in point. These wild grazing animals first came into the city during the region's longest and driest recorded drought. Finding more nourishment in the urban grassland reserves and crown lands, than in the surrounding over-grazed sheep country, they became permanent urban dwellers.

As an addendum to the charming aspects of the children's collective learning with the kangaroos, the research period ended with a grisly discovery. On one of their final walks, we came across the decomposing body of one of the kangaroos. She had been hit by a passing car. Through this confronting experience[3], the children also learned more about the less-than-utopian circumstances of these kangaroos' lives and began to get a sense of the multi-faceted ways in which their own lives are entangled with those of the kangaroos. Realizing that the kangaroos were effectively trapped on the campus, as it is surrounded by major highways, and recalling some of their own close kangaroo road encounters in the family car, they recognised the daily mortal risks faced by their companion kangaroos. This recognition extended the children's attunement to what it might be like to live in a kangaroo's body, in unexpected and complex ways.

Working out how to live together as well as possible in anthropogenically altered and damaged common worlds requires a relational ethics of respect and perseverance – or what Haraway (2008) refers to as an ethics of 'staying with the trouble'. In this kind of ethics, the point is not to seek a final and heroic human-lead solution (which is usually also an ultimately human-centric solution), but to pursue the more modest but still challenging goal of learning how to cohabit with difference in ways that allow all species to 'flourish' (Haraway 2008, 301). To do this we need to become more sensitised to the ways in which we both affect and are affected by other species. One of the most productive ways of sharpening our learning in the interspecies contact zone is to engage with the ethical dilemmas posed by inconvenient and disconcerting cohabitations. This is exactly what Fikile Nxumalo and Veronica Pacini-Ketchabaw are doing in their article in this special issue, and in an earlier coauthored article (Pacini-Ketchabaw and Nxumalo 2015), in which they reflect upon the troubles thrown up by cohabiting children and raccoons in a British Columbian childcare centre. They begin by describing the disconcertment caused by 'unruly' raccoons, whose very presence in the child-care centre contravenes the mandatory licensing regulations to ensure a safe environment for children, who defy all human efforts to keep them out of the playground and who simultaneously charm the children with their endearing cheeky faces and uncannily human-like behaviours. Modeling how we can learn with other animals, through acknowledging their agency and the ways in which they affect us, Pacini-Ketchabaw and Nxumalo reflect upon the raccoons' multifaceted boundary-blurring behaviours as revealing the futility and absurdity of human efforts to demarcate and maintain separate domains – one for us and one for all other animals. They conclude that the wily and beguiling raccoons have a lot to teach us about the ultimate impossibilities of enacting the delusionary nature/culture divide.

Conclusion: becoming worldly as a form of modest recuperation

In this article I have argued that the business-as-usual of environmental stewardship pedagogies, framed by twentieth century humanist change-agency educational discourses, is no longer enough to address the complex imbroglio of twenty-first century human-environmental challenges. The very notion of environmental stewardship is out of step with concurrent moves within environmental education to promote an eco-centric, rather than an anthropocentric view of the environment and indeed, to challenge the nature/culture divide. Even more problematic than this internal inconsistency, environmental and sustainability stewardship approaches remain disconnected from and thus uninformed by critical debates about the Anthropocene naming event that are proliferating in the broader social sciences and humanities. Picking up on the fact that Anthropocene scientists have amassed evidence that human and natural forces, fates and futures are inextricably entwined, these broader debates highlight that it is now time for social science and humanities scholars to move beyond the entrenched humanist paradigms that frame our disciplines and which are still premised upon our separation from (and superiority to) the natural world. Albeit unconsciously (as the mainstream of our discipline remains far removed from such discussions), education's foundational developmental learning theories are still firmly invested in perpetuating this separation.

Feminist critical Anthropocene scholars (for instance Gibson, Rose, and Fincher 2015; Haraway 2015; Stengers 2015a and 2015b; Tsing 2015), whose work has highly influenced the Common Worlds Childhoods Research Collective, are urging us to seize this naming event as an opportunity to redouble efforts to make a complete paradigm shift. They call upon us to make a radical break with humanism's established disorders and to risk experimenting with new mode of collective thinking, or thinking with more-than-human others. The examples I have provided above from recent common worlds research, are attempting to make this radical break in a pedagogical as well as an onto-epistemological sense, by experimenting with modes of collective (more-than-human) thinking and learning with rather than individual (human) thinking and learning about.

Instead of seeking to become better humans by continuing to believe that we are destined to act (alone) on behalf of the world, the common worlds' response to the Anthropocene is quite simply to keep working at ways of become more worldly through focusing upon our entangled relations with the more-than-human world. This is a much more modest response than the ultimately human-centric impulse to break away and 'save' the world. It is a collective or commoning response that refuses human exceptionalism. It is a low-key, ordinary, everyday kind of response that values and trusts the generative and recuperative powers of small and seemingly insignificant wordly relations infinitely more than it does the heroic tropes of human rescue and salvation narratives. These are the kinds of non-divisive relations that many young children already have with the world. They are full of small achievements. We can learn with them.

Notes

1. Liberalist discourses that promote young children as environmental stewards often reiterate developmentalism's essentialist and totalising assumptions about childhood. Blind to their own normative white middle-class centrisms, they posit (all) children's nature experiences as the developmental route towards good environmental citizenship. They rarely acknowledge that childhoods are highly differentiated, as are children's experiences of and cultural meaning making about the nonhuman world.
2. For a detailed explanation of this multispecies ethnographic method, its specific challenges and affordances, see Pacini-Ketchabaw, Taylor, and Blaise 2016.
3. For a detailed account of the children's responses to the dead kangaroo, see Taylor & Pacini-Ketchabaw (2016).

Disclosure statement

No potential conflict of interest was reported by the author.

ORCID

Affrica Taylor (iD) http://orcid.org/0000-0002-9222-1278

References

Alaimo, S. 2010. *Bodily Natures: Science, Environment and the Material Self*. Bloomington: Indiana University Press.
Blanchard, P. B., and T. K. Buchanan. 2011. "Environmental Stewardship in Early Childhood." *Childhood Education* 87 (4): 232–238.
Chakrabarty, D. 2009. "The Climate of History: Four Theses." *Critical Inquiry* 35: 197–222.
Chawla, L. 2009. "Growing up Green: Becoming an Agent of Care for the Natural World." *The Journal of Developmental Processes* 4 (1): 6–23.
Chawla, L., and M. Rivkin. 2014. "ECEfS in the USA." In *Research in Early Childhood Education for Sustainability: International Perspectives and Provocations*, edited by J. M. Davies and S. Elliott, 248–265. London: Routledge.
Colebrook, C. 2011. "Earth Felt the Wound: The Affective Divide." *Journal for Politics, Gender and Culture* 8 (1): 45–58.
Common World Childhoods Research Collective. 2016. *Common World Childhoods Research Collective Website*. www.commonworlds.net.
Crist, E. 2013. "Provocation: On the Poverty of Our Nomenclature." *Environmental Humanities* 3: 129–147.
Cronon, W. 1998. "The Trouble with Wilderness, or, Getting back to the Wrong Nature". In *The Great New Wilderness Debate*, edited by J. B. Callicott and M. P. Nelson, 471–499. Athens: University of Georgia Press.
Crutzen, P. 2002. "Geology of Mankind." *Nature* 415: 23. doi:10.1038/415023a.
Cutter-Mackenzie, A., S. Edwards, D. Moore, and W. Boyd 2014. *Young Children's Play and Environmental Education in Early Childhood Education*. Amsterdam: Springer.
Dahlberg, G., and P. Moss. 2005. *Ethics and Politics in Early Childhood Education*. London: RoutledgeFalmer.
Davis, J., ed. 2015. *Young Children and the Environment Early Education for Sustainability*. 2nd ed. New York: Cambridge University Press.
Davis, J. M., and S. Elliott, eds. 2014. *Research in Early Childhood Education for Sustainability: International Perspectives and Provocations*. London: Routledge.
Deleuze, G., and F. Guattari. 1987. *A Thousand Plateaus: Capitalism and Schizophrenia*. Minneapolis, MN: University of Minnesota Press.
Duhn, I. 2012. "Places for Pedagogies, Pedagogies for Places." *Contemporary Issues in Early Childhood Education* 13 (2): 99–107.

Edwards, S. 2014a. "Environmental Education and Pedagogical Play in Early Childhood Education." In *Young Children's Play and Environmental Education in Early Childhood Education*, edited by A. Cutter-Mackenzie, S. Edwards, D. Moore, and W. Boyd, 25–37. Amsterdam: Springer.

Edwards, S. 2014b. "A Challenge Reconsidered: Play-Based Learning in Early Childhood Environmental Education." In *Young Children's Play and Environmental Education in Early Childhood Education*, edited by A. Cutter-Mackenzie, S. Edwards, D. Moore, and W. Boyd, 75–81. Springer.

Elliott, S., and T. Young. 2015. "Nature by Default in Early Childhood Environmental Education?" *Australian Journal of Environmental Education* 32 (1): 1–8.

Fawcett, L. 2002. "Children's Wild Animal Stories: Questioning Interspecies Bonds." *Canadian Journal of Environmental Education* 7 (2): 125–139.

Fraser, C. 2009. *Rewilding the World: Dispatches from the Conservation Revolution*. New York: Metropolitan Books.

Gannon, S. 2015. "Saving Squawk? Animal and Human Entanglements at the Edge of the Lagoon." *Environmental Education Research* 23 (1): 91–110. doi:10.1080/13504622.2015.1101752.

Gibson, K., D. B. Rose, and R. Fincher, eds. 2015. *An Ethics for the Anthropocene*. New York: Punctum Books.

Greenwood, D. 2014. "Culture, Environment and Education in the Anthropocene". In *Assessing Schools for Generation R (Responsibility)*, edited by M. P. Mueller, D. J. Tippins, and A. J. Stewart, 279–292, Amsterdam: Springer.

Hamilton, C. 2014. "The delusion of the 'Good Anthropocene': Reply to Andrew Revkin", June 17. http://clivehamilton.com/the-delusion-of-the-good-anthropocene-reply-to-andrew-revkin/

Hamilton, C. 2015. "Human Destiny in the Anthropocene". In *Anthropocene and the Global Environmental Crisis: Rethinking Modernity in a New Epoch*, edited by C. Hamilton, C. Bonneuil, and F. Gemmene, 32–43, London: Routledge.

Haraway, D. 2008. *When Species Meet*. Minneapolis: University of Minnesota Press.

Haraway, D. 2015. "Anthropocene, Capitalocene, Plantationocene, Chthulucene: Making Kin." *Environmental Humanities* 6: 159–165.

Haraway, D., N. Ishikawa, S. Gilbert, K. R. Olwig, A. L. Tsing, and N. Bubandt. 2015. "Anthropologists Are Talking – About the Anthropocene." *Ethnos* 81 (3): 535–564. doi:10.1080/00141844.2015.110583.

Hird, M. J. 2013. "Waste, Landfills, and an Environmental Ethic of Vulnerability." *Ethics and the Environment* 18 (1): 105–124.

Hodgins, D. 2015. "Wanderings with Waste." *Canadian Children* 40 (2): 88–100.

Huckle, J. 1996. "Education for Sustainable Citizenship: An Emerging Focus for Education for Sustainability". In *Education for Sustainability*, edited by J. Huckle, and S. Sterling, 228–244, Abingdon: Earthscan for Routledge.

Instone, L. and Taylor, A. 2015. "Thinking about Inheritance through the Figure of the Anthropocene, from the Antipodes and in the Presence of Others". *Environmental Humanities* 7:133–150.http://environmentalhumanities.org/arch/vol7/7.7.pdf.

Kopnina, H. 2012. "Education for Sustainable Development (ESD): the Turn Away from 'Environment' in Environmental Education?" *Environmental Education Research* 18 (5): 699–717.

Kyburz-Graber, R. 2013. "Socioecological Approaches to Environmental Education and Research". In *International Handbook of Research in Environmental Education*, edited by R. B. Stevenson, M. Brody, J. Dillon, and A. E. J. Wals, 23–32, New York: Routledge.

Langton, M. 1996. "What Do We Mean by Wilderness? Wilderness and *Terra Nullius* in Australian Art." *The Sydney Papers* 8 (1): 10–31.

Latour, B. 2004. *The Politics of Nature: How to Bring the Sciences into Democracy*. Translated by C. Porter. Cambridge, MA: Harvard University Press.

Latour, B. 2014. "Agency at the Time of the Anthropocene." *New Literary History* 45: 1–18.

LLoro-Bidart, T. 2016. "A Political Ecology of Education in/for the Anthropocene." *Environment and Society: Advances in Research* 6: 128–148.

Lorimer, J. 2015. *Wildlife in the Anthropocene: Conservation after Nature*. Minneapolis, MN: University of Minnesota Press.

Malone, K. 2015. "Posthumanist Approaches to Theorising Children's Human-Nature Relations." In *Space, Place and Environment, Volume 3, Geographies of Children and Young People Series*, edited by K. Nairn and P. Kraftl, 1–22. London: Springer.

Malone, K., T. Gray, and S. Truong, eds. 2017. *Reimagining Sustainability Education in Precarious Times*. Amsterdam: Springer.

Nxumalo, F. 2015. "Forest Stories: Restorying Encounters with 'Natural' Places in Early Childhood Education." In *Unsettling the Colonial Places and Spaces of Early Childhood Education*, edited by V. Pacini-Ketchabaw and A. Taylor, 21–42. New York: Routledge.

Oakley, J. 2011. "Animality and Environmental Education: Towards an Interspecies Paradigm." *Canadian Environmental Education* 16: 8–13.

Oakley, J., G. Watson, C. Russell, A. Cutter-Mackenzie, L. Fawcett, G. Kuhl, J. Russell, M. van der Waal, and T. Warkentin. 2010. "Animal Encounters in Environmental Education Research: Responding to the 'Question of the Animal.'" *Canadian Journal of Environmental Education* 15: 86–102.

Pacini-Ketchabaw, V. 2013. "Frictions in Forest Pedagogies: Common Worlds in Settler Colonial Spaces." *Global Studies of Childhood* 3 (4): 355–365.

Pacini-Ketchabaw, V., S. Kind, and L. Kocher. 2016. *Encounters with Materials in Early Childhood Education*. New York: Routledge.

Pacini-Ketchabaw, V., and F. Nxumalo. 2015. "Unruly Raccoons and Troubled Educators: Nature/Culture Divides in a Childcare Centre." *Environmental Humanities* 7: 151–168.

Pacini-Ketchabaw, V., and A. Taylor 2015. "Unsettling Pedagogies through Common World Encounters: Grappling with (Post) Colonial Legacies in Canadian Forests and Australian Bushlands". In *Unsettling the Colonialist Places and Spaces of Early Childhood Education*, edited by V. Pacini-Ketchabaw and A. Taylor, 43–62. New York: Routledge.

Pacini-Ketchabaw, V., and A.Taylor. 2016. "Common World Childhoods". In H. Montgomery (Ed.) *Oxford Bibliographies in Childhood Studies*. New York: Oxford University Press. doi:10.1093/OBO/9780199791231-0174

Pacini-Ketchabaw, V., A. Taylor, and M. Blaise 2016. "De-Centring the Human in Multispecies Ethnographies". In *Posthuman Research Practices in Education*, edited by C. Taylor and C. Hughes, 149–167. Houndmills: Palgrave Macmillan.

Prout, A., and A. James, eds. 1990. *Constructing and Reconstructing Childhood: Contemporary Issues in the Sociological Study of Childhood*. London: RoutledgeFalmer.

Rautio, P. 2013a. "Being Nature: Interspecies Articulation as a Species-Specific Practice of Relating to Environment." *Environmental Education Research* 19 (4): 445–457.

Rautio, P. 2013b. "Children Who Carry Stones in Their Pockets: On Autotelic Material Practices in Everyday Life." *Children's Geographies* 11 (4): 394–408.

Rautio, P., and P. Jokinen. 2016. "Children's Relations to the More-than-Human World beyond Developmental Views." In *Geographies of Children and Young People: Play, Recreation, Health, and Well Being*, edited by T. Skelton, J. Horton, and B. Evans, 35–49. Singapore: Springer.

Rose, D. B., T. van Dooren, M. Chrulew, S. Cooke, M. Kearnes, and E. O'Gorman. 2012. "Thinking through the Environment, Unsettling the Humanities." *Environmental Humanities* 1: 1–5.

Somerville, M. 2017. "The Anthropocene's Call to Educational Research." In *Reimagining Sustainability Education in Precarious times*, edited by K. Maloney, T. Gray, and S. Truong, 17–28. Amsterdam: Springer.

Somerville, M., and M. Green. 2015. *Children, Place and Sustainability*. New York: Palgrave Macmillan.

Steffen, W., W. Broadgate, L. Deutsch, O. Gaffney, and C. Ludwig. 2015. "The Trajectory of the Anthropocene: The Great Acceleration." *The Anthropocene Review* 2(1):81–98.

Steffen, W., P. Crutzen, and J. R. McNeill 2007. "The Anthropocene: Are Humans Now Overwhelming the Great Forces of Nature?" *AMBIO: A Journal of the Human Environment* 36 (8): 614–621.

Stengers, I. 2012. "Cosmopolitics: Learning to Think with Science, Peoples and Natures". Public lecture, St. Marys University, Halifax, Canada, March 5. https://www.youtube.com/watch?v=1l0ipr61SI8

Stengers, I. 2013. "Matters of Cosmopolitics: On the Provocations of Gaia (in Conversation with Heather Davis and Etienne Turpin)." In *Architecture in the Anthropocene: Encounters among Design, Deep Time, Science and Philosophy*, edited by E. Turpin, 171–182. Michigan: Open Humanities Press.

Stengers, I. 2015a. *In Catastrophic times: Resisting the Coming Barbarism*. Paris: Open University Press.

Stengers, I. 2015b. "Accepting the Reality of Gaia: A Fundamental Shift?" In *The Anthropocene and the Global Environmental Crisis: Rethinking Modernity in a New Epoch*, edited by C. Hamilton, C. Bonneuil, and F. Gemmene, 134–144, London: Routledge.

Stevenson, R. B., M. Brody, J. Dillon, and A. E. J. Wals, eds. 2013. *International Handbook of Research on Environmental Education*. New York: Routledge (and AERA).

Taylor, A. 2013. *Reconfiguring the Natures of Childhood*. London: Routledge.

Taylor, A., M. Blaise, and M. Giugni. 2013. "Haraway's 'Bag Lady Story-Telling': Relocating Childhood and Learning within a 'Post-Human Landscape'." *Discourse: Studies in the Cultural Politics of Education* 34 (1): 48–62.

Taylor, A. 2017. "Romancing or Reconfiguring Nature? Towards Common Worlds Pedagogies." In *Reimagining Sustainability Education in Precarious Times*, edited by K. Maloney, T. Gray and S. Truong, 61–75. Amsterdam: Springer.

Taylor, A., and M. Giugni. 2012. "Common Worlds: Reconceptualising Inclusion in Early Childhood Communities." *Contemporary Issues in Early Childhood* 13 (2): 108–119.

Taylor, A., and V. Pacini-Ketchabaw. 2015. "Learning with Children, Ants, and Worms in the Anthropocene: Towards a Common World Pedagogy of Multispecies Vulnerability." *Pedagogy, Culture, Society* 23 (4): 507–529.

Taylor, A., and V. Pacini-Ketchabaw. 2016. "Kids, Roos, and Raccoons: Awkward Encounters and Mixed Affects." *Children's Geographies* 15 (2): 131–145. doi:10.1080/14733285.2016.1199849.

Tsing, A. L. 2015. *The Mushroom at the End of the World: On the Possibility of Life in Capitalist Ruins*. Princeton, NJ: Princeton University Press.

Tuck, E., M. McKenzie, and K. McCoy, eds. 2014. "Land Education: Indigenous, Post-Colonial and Decolonizing Perspectives on Place and Environmental Education Research." Special Issue, *Environmental Education Research* 20 (1): 1–23.

Wattchow, B., R. Jeanes, L. Alfrey, T. Brown, A. Cutter-Mackenzie, and J. O'Conner, eds. 2014. *The Socioecological Educator: A 21st Century Renewal of Physical, Health, Environmental and Outdoor Education*. Amsterdam: Springer.

Wilson, E. O. 1993. "Biophilia and the Conservation Ethic." In *The Biophilia Hypothesis*, edited by S. R. Kellert and E. O. Wilson. Washington DC: Shearwater/Island Press.

Wilson, R. A. 2011. "Becoming Whole: Developing an Ecological Identity." *Wonder*, May/June edition of Newsletter of Nature Action Collaborative for Children. www.ccie-media.s3.amazonnews.com/nacc/wonder_may11.pdf.

Yusoff, K. 2013. "Geologic Life: Prehistory, Climate, Futures in the Anthropocene." *Environment and Planning D: Society and Space* 31 (5): 779–795. http://www.restaurant-relae.dk/en/mad/.

Yusoff, K. 2016. "Anthropogenesis: Origins and Endings in the Anthropocene." *Theory, Culture and Society* 33 (2): 3–28.

Reconfiguring urban environmental education with 'shitgull' and a 'shop'

Pauliina Rautio, Riikka Hohti, Riitta-Marja Leinonen and Tuure Tammi

ABSTRACT

The worry over urban children having lost their connection to nature is most often addressed with either initiatives of reinserting the 'child back to nature' or with evidence aiming to prove that the worry is unfounded to begin with. Neither approach furthers our understanding of child–nature relations as continuing transformation of both 'child' ('human') and 'nature'. The objective of this paper is to redirect attention from evaluating *connectedness* of two separate units to mapping *mutual emergence* of children and their surroundings in relation to each other. The question asked is: *Of what kind is environmental education beyond connectedness of 'child' and 'nature'?* The aligned theoretical approach, (critical) posthumanism, will help us to elaborate a premise for environmental education according to which humans and their nonhuman surroundings do not exist as independent of each other. The empirically grounded events discussed in this paper are named 'shitgulls' and 'shops'. These events map mutual emergence of child and nature, evidencing the need for environmental education to understand itself as a relational phenomenon.

The persistent worry over urban children having lost their connection to nature is most often addressed with either initiative of reinserting the 'child back to nature' (Malone 2015b) or arguments stating that the worry is unfounded to begin with. Urban children *are* connected, and that *there is nature* available for exploration: in the cracks and crevices of cement, in the footprints of foxes and city rabbits. The underlying implications of both responses often remain unchallenged however: that humans are able to become detached from nature in the first place. That 'human' and 'nature' can be predetermined as to what they are (rather than always only defined through the relation under study). And so the anthropocentric predicament remains, contrary to what is often the intended argument: humans are not considered as part of nature (e.g. Clarke and Mcphie 2014; Rautio 2014; Malone 2015a, 2015b).

The focus of environmental education (research) on two separate entities: 'child' (or 'human') and 'nature', manifests in elaborate means and scales to evaluate the 'connectedness' of the two (e.g. Schultz 2002; Ernst and Theimer 2011; Liefländer et al. 2013; Bruni et al. 2017). Resulting is an almost medicalized concern for the supposed lack of children's connectedness with nature most famously diagnosed as the 'nature-deficit-disorder' (Louv 2005), or 'environmental developmental amnesia' (Kahn 2002). These medico-psychological terms construct the challenge of environmental education as that of addressing individual deficits and crafting solutions that target individual deficits – that is, inventing cures. While no doubt informative, measuring connectedness is not the only way of understanding and addressing children's relations with their environments. And in light of the focus of the special issue at hand, a view of urban children as deprived and lacking of connectedness to nature leaves a majority of the world's children and young people portrayed as victims or patients in need of rescue or remedy. As Claudia Diaz shows in her article in this special issue, children's relations to and conceptions of 'urban' or 'nature' develop also beyond their immediate surroundings. Taking up Donna Haraway's insistence that 'it matters what stories make worlds' (Haraway 2016, 12) this paper presents another story of urban children's nature relations.

The objective of this paper is to redirect attention from evaluating *connectedness* of two separate units to mapping *mutual emergence* of children and their surroundings in relation to each other. The question asked is: *Of what kind is environmental education beyond connectedness of 'child' and 'nature'?* The aligned theoretical approach, (critical) posthumanism, will help us to elaborate a premise for environmental education according to which humans and their nonhuman surroundings do not exist as independent of each other (Pedersen 2010; Snaza and Weaver 2014; Rautio 2014; Malone 2015a). The empirical anchoring of this paper is in a study in which children's relations to nature are researched as mutually emerging – both conceptually and as of certain kind in practice – through their daily engagements. These engagements are framed as *child-within-nature events*, and two are discussed in detail.

The study that frames this paper took place in an urban setting of the North of Finland in the winter of 2013 with 12 participating children of 6–8 years old. The researcher accompanied the participating children to diverse locations around the largest city in the North of Finland (c. 200 000 inhabitants) engaging in multimodal ethnographic mapping of mutually emerging child-within-nature events. Two resulting events discussed in this paper in depth are the ones named 'shitgulls' and 'shops'. The first event took place in a main landfill and recycling site of the city and surrounding areas. In addition to me and the children 'the urban' and arguably 'the nature' took part and generated each other as green containers, crisscrossing pathways, roads, compost piles, electronics, tyres, cardboard, glass, descending darkness, fluorescent lights, extreme cold, birds and rats and heavy machinery moving around discarded materials from 300 000 people. In the second event, a local suburban grocery shop with surrounding areas such as the parking lot, the bike paths leading to it, and the adjacent snow piles (materialisations of urban transportation infrastructure), the even more extreme cold and darkness, and the warm commercial indoor space were the mutually generative elements that were identified as a child-within-nature event.

The event of the shitgull challenges persistent tendencies of viewing both children and their connection with nature as inherently good and desirable (see Tipper 2011). When an 8-year old child proclaims 'That's a shitgull. They eat shit. They ought to be shot', simplified accounts of children's authentic affinity towards nature or animals in specific are challenged. Instead, the child–within–nature becomes a raw existence based on, in this case, mutual disaffect and avoidance. The event of the shop challenges the discursive definition of 'nature' as a separate 'natural surrounding'. Children's immediate and bodily existence in the wintery nature of Northern Finland calls for regulation of temperatures: they move inside when they get too cold to play outside, making the warm cornershop an essential part of their relation to 'nature'. In this sense, the child–nature becomes something more entangled than a separate backdrop to human actions.

To enable children to develop relations to their urban environments, to realise their own embeddedness in nature, their evolving with all that they encounter, we need to know all that's part of these relations, not just the isolated human and/or nonhuman agents. We need to know how a seagull becomes a 'shitgull'. Or better: how 'shitgull' emerges from complex intra-actions and how *only after* the uttered 'shitgull' can we dissect the event and talk retrospectively about a 'child' and a 'seagull' as some of the agents in the event.

Environmental education beyond humanism

To be able to go beyond the initiatives of simply inserting the child back into nature requires first recognising the grounding anthropocentric philosophies of education at large and environmental education in specific. Lodged for long within processes and practices of humanistic education, environmental education needs to address its philosophical traditions (Russell 2005). What is the 'human' in 'becoming human' – both as a species and a social and political category? And how has it changed over time? How is it possible to make a claim that 'children have lost their connection to nature' if not from the understanding that 'children' and 'nature' are ontologically separate rather than mutually embedded categories. The following review in this paper is a general overview of the work done so far, intended to pave way for the following discussion of child-within-nature configurations.

An orientation for thinking about education beyond anthropocentrism is the diverse and increasingly popular philosophical approach of posthumanism (see e.g. Snaza et al. 2014). The ideas of posthumanism – with the 'post' referring to a fundamental ontological and epistemological shift rather than to a chronological 'coming after' – are not new. Albeit debated (see e.g. Gane 2006), the roots and rhizomes can be taken to date back and go somewhat far, especially when looking at the ideas rather than tracing the term 'posthuman(ism)'. A key point of reference is Bergson's ([1903] 2007) metaphysics and the idea of intuition as knowing something from within. Intuition, for Bergson, is not a supernatural effort, rather the effort of experiencing ourselves as more or other than just human. Perhaps easier to access, however, is Pickering's (2005) discussion of posthumanism in the interface of natural sciences and social sciences between which there is a neat labor division: natural sciences focus on things, social sciences on people/meanings. Pickering argues that this dualism produces two sets of units of analysis, either objects or subjects but not both – save their interaction perhaps. What is left out is the interdependence between objects and subjects, or matter and meaning. Pickering argues that this presumption of the two possible units of analysis is not compulsory. And that once the posthumanist unit of analysis – interdependence – is taken up, a *genuine object of inquiry* is conjured (Pickering 2005, 34). A unit of analysis which is not about objects/things/matter *or* people/meanings but the heterogeneous assemblage comprising both, the 'dance of human and nonhuman agency' (Pickering 1995). This is precisely what is discussed and described in this paper: the child-within-nature configurations as objects of inquiry because of the complex interdependencies entailed.

Until recently education has remained one of the last fortresses of anthropocentrism keeping 'child' and 'nature' categorically apart, achieving at best merely a *symbolic* decentering of the human subject (Pedersen 2010, 242, 243). The real shift would be to acknowledge how nonhuman animals and environments are already part of our human selves: partners in our 'identity-forming relationships' (Pedersen 2010, 243) and participants in our 'network of learners' (Rocheleau 2015, 57; Lloro-Bidart 2016, 6) and thus at the core of *being human* (cf 'becoming human' as moving further from 'being animal'). These are familiar claims long repeated for instance within human–animal studies or cultural studies at large (see e.g. Weil 2010) and originating largely from classics such as Donna Haraway's (2003) *The Companion Species Manifesto* or *When Species Meet* (2008). In turn Despret (2004, 131; see also 2016), also influential yet less credited, speaks of 'anthropo-zoo-genetic practices' referring to situations where species domesticate each other creating new articulations of 'with-ness'.[1]

More and more scholars of education are discussing the humanistic underpinning of education: how it emerged, how it manifests in discourses and practices and with what consequences. Pedersen (2010, 2015) among others (e.g. Weaver 2010; Snaza et al. 2014; Rice and Rud 2016) traces the intellectual roots of education to the era of Enlightenment and concludes that the idea of education is, and has been since, symbolised by the questions of what it means to be *human* and to function in a *human society* (Pedersen 2015, 51; see also McKay 2005, 218). Livingston and Puar (2011, 8) conceptualise education as a bio-political process by which humans become a species: a series of didactic investments in the *human form* and as such an anthropogenesis or the historical articulation between human and animal (see Agamben 2004, 79).

This overview of the various conceptualisations evoked has shown that intellectual conditions for thinking beyond the 'child' and 'nature' dualism exist, and have existed for a while (e.g. Russell, Sarick, and Kenelly 2002). Empirical educational research and more specific theorizations are beginning to emerge as well (Rice and Rud 2016) exemplifying the next step after acknowledging anthropocentrism: the active unlearning of it (Lupinacci and Happel-Parkins 2016). From Andrew Pickering's (2005) study of interdependencies of Asian Eels, global warming and politics to Lloro-Bidart's (2016) feminist posthumanist political ecology of education for theorizing human–avian relationships the identifying of anthropocentrism has turned into transcending of it. What seems to be shared in these movements is a pursuit of the 'affective intensities of what bodies or things could do in the act of becoming-with' (Snaza et al. 2014, 47), that is, an attunement to real-life experiences, embodied, emotional and affective ways of knowing (Lloro-Bidart 2016, 2). Through this pursuit, a new politics becomes possible; one that emerges from realising the limits of humanist versions of democracy – the political implications of the very category of 'human' – and the possibility of 'political forms that are not narrowly restricted to humans' (Snaza et al. 2014, 49) and their possessed interests.

From this perspective, the polis – the community in which relations emerge as problems – can be treated as more-than-human, situated in the affectual streams of everyday life. And it follows that simple deficit theories that presuppose the separate and predetermined 'child' and 'nature' are no longer the only plausible explanation.

Multimodal ethnographic wandering

If we have it that practices of education locate in multiple and messy relations between humans and all that surround them (Fenwick, Edwards, and Sawchuck 2011, 709; Sørensen 2009, 170, 171) then the research of these relations could very well reflect and explore rather than seek to clean up this messiness (Law 2004). And maybe even further: not only accommodating existing mess but also deliberate practices of making accidents and mess abound in research could be taken up in educational research. Following Pickering's (2005) insight that a genuine posthumanist object of inquiry are the interdependencies of objects and subjects, this kind of methodological approach would target precisely those entanglements. So what MacLure (2010, 730) describes as 'interruptive method'[2] was understood in the study as method which enables exploration and experimentation of interdependency rather than analysis of subjects or objects, or even their interaction. Interdependencies and affectual assemblages were brought to light by inducing research situations with deliberate interruptions and difficulties (see Niccolini and Pindyck 2015). During the two-year study the participating children, the locations of our meetings and the things brought along to take part in our meetings were not always decided, planned or even known beforehand by me (*Author A*, 2013). The difficulties induced for data documentation were of particular importance to this paper at hand.

The study took place in Northern Finland (years 2012–2013) with 12 participating children of 6–8 years old. The wider focus was to explore the intra-actions between children and matter in informal urban settings. The distinction between the more familiar 'interactive' and 'intra-active' is that the former refers to two *independent entities* taking turns in affecting each other ('child' and 'nature'), whereas in the latter two interdependent entities come into being through simultaneous activity, continually defining each other (see especially Barad 2007).

The research meetings took place outside in various public or semi-public urban locations. To follow the notion of interrupting a conventional focus on objects/subjects and let interdependencies surface I deliberately failed for example at 'data documentation'. I knew that in below minus 20 °C temperatures my recorder would not work, nor could I write down notes or photograph well. I had to switch to experimenting with what multisensory ethnography (Pink 2009) might be in this context, lingering on various expressive events – combinations of words, gestures, smells, heights, temperatures, darkness, ice, dirt, and sounds. I made headnotes which I later turned into fieldnotes with the help of scribbled notes on papers, the few minutes of soundscapes and recorded talk that my phone would record before it froze (sometimes I kept the phone under my coat, next to my skin and it warmed up again for a few more minutes), and any random objects the children might have picked up and handed to me. But mostly I relied on intense observation and mental notes. This distorted linear time and defied coherent accounts.

By making research practices difficult and messy I had no choice but to tune into the details that caught my attention and work with them. To create temporary cuts in the flat, ever reorganizing plateau. This is an approach to qualitative inquiry and to data that Brinkmann (2014) calls abductive or breakdown-driven, and MacLure (2013) characterises as openness to surprises and to the mutuality of researcher and data constituting each other. Both challenge the idea that data is collected; Brinkmann (2014, 722) suggests that inquiry arises out of a surprising situation rather than in relation to collected data:

> To sum up, the abductive approach presents research as part of the life process, as what we do in situations of breakdown that inevitably arise in life's situations – big or small. There is no talk here of collecting data or framing them theoretically, but rather of navigating existential, moral, and political situations as individuals and collectives.

'That's a Shitgull' said one child pointing at a Seagull flying above our heads at the landfill site we were visiting. 'My dad says they're shit birds'. The heard, overheard, seen and felt begun to cluster around the inappropriately identified bird: 'Shit birds deserve to die', 'I am the only girl', 'If I fell there would I die?' In the extreme cold the visible warm breath arising from the child's mouth as he voiced 'shit' or 'death' all became as if data to think with. The all-dreaded binaries surfaced: life/death, human/animal, worthy/shit, boy/girl, good/bad. It was a politicizing moment as it invoked binaries and order between things that were telling of the cultural conventions that gave rise to understanding and working with the event. What possibilities for understanding are currently open to us?

The discussion in this paper leans explicitly on two events. Following Denzin (2001, 63) the events are trusted to work as sources for more general discussion as evidence of the cultural conditions that make them and the interpretations they beg possible. With these two instances we aim to show both the available conventional readings and understandings of them as well as to provide alternative, less coherent and more complex readings – because it can be done, because the philosophical basis of (environmental) education does not have to be explicitly anthropocentric anymore.

Shitgull as a child–within–nature configuration

Until recently when reviewing literature on child–animal relations(hips) the dual categories of 'child' and 'animal' seemed to have generated research of two 'others' from the viewpoint of an onlooker, an adult citizen who herself belonged to neither category (Melson 2005; Mason and Tipper 2008; Tipper 2011). Majority of these studies were about the psychological/developmental aspects of children's engagements with animals, most often pets or so called therapy animals (e.g. Alach 2003; Thompson and Gullone 2003; Melson 2005; Friesen 2010; Prokop and Tunnicliffe 2010; McCardle et al. 2011; Myers 2013). While human–animal studies in general has a long history in accounting for mutuality and com-plexity of human–animal relations (DeMello 2012), research of children and animals in particular has remained focused on categories and developmental stages (Taylor, Blaise, and Giugni 2013). Research including children has, until recently, emphasised the 'good or bad' tug of war, that is focused on the influence animals have on children (McCardle et al. 2011) rather than the experienced relationship from the viewpoint of the individual children and animals (Tipper 2011). Without knowledge of the experienced, everyday and often contradictory nature of child–animal relations these bonds often end up being either romanticised or valued only for their utility.

Studies of child–animal relations(hips) designed based on the idea that there are predetermined categories of 'child' and 'animal' (or 'pet', 'dog', 'cat') are still not hard to come by. These cultural categories are alive and well, also in research. Comparing the socio-emotional characteristics of children pet owners and children without pets, Vlasta Vizek Vidović and team address the *benefits* of a certain *type of pet* to *different children's* socio-emotional *development* (Vidović, Štetić, and Bratko 1999, 211):

> In order to answer the main research question, several analyses of variance (gender by grade by pet ownership) were computed for each criterion of socio-emotional development. Significant main effects were obtained for empathy, prosocial orientation and pet attachment, with dog owners being more empathic and prosocially oriented than non-owners, and dog owners and cat owners being more attached to their pets than owners of other kinds of pets.

If from the onset the objective is to study a bond between two species, say humans and dogs – rather than a bond between two particular beings who share a lifeworld, evolving and cohabitating as 'Annie' and 'Sparky' – one might ask in which way have these child–animal relations studies been about 'rela-tions(hips)'? The level of generalisation loses not only animal individuals but human child individuals and cannot get to the core of how a bond emerges between two beings of different species (Pedersen 2010, 2015; Tipper 2011) – of how interdependence emerges. If Annie loves Sprky, feels safe and comforted in their shared play, and sleeps better with Sparky by her bed we cannot conclude that Annie will be the same with Fido, or Rex. Annie-Sparky is not the same relationship as Annie-Fido, or Annie-Rex.

In the past 10–15 years, however, what has been reported as an 'animal turn' (Weil 2010) has taken place. Calls and advances for outlining the ways in which children are both constituted by and learn with their significant more-than-human others are made among other disciplines also in education (Watson 2006; Oliver 2009; Taylor, Blaise, and Giugni 2013; Nelson, Coon, and Chadwick 2015; Taylor and Pacini-Ketchabaw 2015). And they are often made in ways that privilege everyday events, affectual and embodied knowing-with beyond species limits and attending to mutual vulnerabilities such as with wasp-bee-mushrooms (Atkinson 2015), or ants, worms and children (Taylor and Pacini-Ketchabaw 2015).

An evident tendency to focus on positive and beneficial child–animal relations leaves out a wealth of undesired, conflicting and even harmful relations (to either the child or the animal, or both) which are nevertheless part of children's and other animals' lives (Cole and Stewart 2014; Rowe 2016; Lewis 2016) and often have to do with death or exploitation of animal individuals (Cole and Stewart 2014). Building on the work of Pacini-Ketchabaw and Nxumalo (2015; see also Taylor and Pacini-Ketchabaw 2016) of awkward and frictional encounters between racoons and educators and Nelson's (2016) figurings of death in early childhood practices the child-within-nature configuration of a Shitgull is discussed as an unruly child–animal relation, indicative of the conventional ways of seeing children and animals with what it clashes. Ways of reading the event include focusing on interdependencies, diverse agential elements of the event and mutual vulnerabilities in the relation.

We had met with the participating children in many places already. This time they had wished to explore the local landfill and recycling centre. They should have worn bright yellow vests but there was nobody to meet us at the reception of the landfill area. As the area is semi-public, and as I had asked for a permission and gotten it in writing, and as we weren't allowed to walk anywhere but the safe pedestrian areas anyway, we proceeded even without the vests. In the absence of the vests – the explicit purpose of which I hadn't been told – I felt uneasy and wanted to limit the children's roaming as they weren't 'tagged'. I had little luck and proceeded to mimic a sheep dog trying to go round the children to keep them together. One of them stayed with me. I knew him from my neighbourhood and I knew of his family as well. His parents had recently separated and his dad had moved far away to another city. They hardly ever saw him anymore. The boy came to visit our house sometimes to look at the injured birds we took care of as part of our wildlife rehabilitation work.

We were standing at the edge of an uncovered compost area with the child. Rest of the children were further away. The area was filled with birds scavenging for food. A Seagull flew over our heads. 'That's a SHITgull!' said the boy pointing at the bird. 'My dad says they're SHIT birds and they ought to be SHOT' he said, looking at me. And there was a child–animal relation with intense emotions and far-reaching personal and political relevance and significance to both the child and the bird. All infused in a brief encounter. The following is an example of the thinking (conventionally called 'analysis') done with and around the event to highlight various interdependencies.

> The shitgull as a child–within–nature configuration was a fleeting bond that was simultaneously enabled by and transgressed the boundaries of 'child' and 'nature'. The bird was looking for food when we interrupted him. Leftover food that had been collected from humans and dumped in the huge open compost. Causing seagulls, ravens, crows and magpies to flock and populate the compost heaps, attracting also rats and smaller rodents. Causing the landfill personnel to put up scarecrows and nets to which the birds would get tangled and hang flapping upside down until their slow death. Because of eating human waste. Shit. The relation between the Seagull and the boy was that of mutual disaffect and avoidance – of categorical, concrete and symbolic *using* of each other. The Seagull for food, the boy for seeking a let out for his feelings and confusion over his dad, it seemed. Without the bird, the boy could not have been an unhappy, angry, death-wishing child – an adamant and intentional non-child. Knowing that I take care of injured birds, plenty of seagulls included, his wish for the bird to be dead, pointed at me, was especially weighty. As the child declined his childness the bird declined his wildness. He survived on human waste and witnessed his fellow gulls dying, tangled in the nets above the compost. The 'shitgull' was an event of ill-being for all involved; yet it was an event of the utmost interdependence. An interdependence gone wrong.

Destruction and death have always been the key in the enacting of boundaries between animals and humans – be they raccoons (Pacini-Ketchabaw and Nxumalo 2015) or seagulls, pigeons, or hawks (Lorimer 2012). Destruction and death have for long been the low-key in the enacting of childhood in early childhood education (Nelson 2016). The child-within-nature configuration of a shitgull gives rise to a troubling reality in which both childhood and seagulhood are mutually enacted and implicated: the gull risks his life, the child his humanity (or animality).

It is this diluting of a clear, identifiable one-to-one, often causal relations through which humanist approaches to viewing the world, to viewing education, can be dismantled. The posthuman condition is by definition an uncertain one (Pepperell 2003), a temporal one (Pickering 2005), and one in which bodies, bodies of discourses and discourses of bodies intersect (Halberstam and Livingstone 1995, 2). There never was a 'child' and a 'Seagull' outside of their relation which included all of the discourses and bodies that made a 'Shitgull'. The unit of analysis and object of inquiry was not 'child + Seagull' but an evolving interdependent assemblage of the 'Shitgull' the process of which we could not have known in advance (Pickering 2005, 34) but which opened up among others a way to address disaffection in child–animal encounters without pathologising the child.

'Shop' as a child–within–nature configuration

When discussing urban children's relations to nature, the necessary first step is to delineate what counts as 'nature'. A long standing and arguably 'western' tradition has been to consider nature as pure, wild, unchanging and as contrary to the continually changing cultures of humans (Lorimer 2012; Granjou 2016). Karen Malone (2015b) reviews and discusses some of the drawbacks of the Child in nature -movement, mainly how the romanticised idea of 'authentic' nature reifies rather than resolves the human–nature split.

From a posthumanist point of view 'nature' and 'child' are not essential a priori beings, rather mutually emerging materially and discursively. The quest is then to define this mutual emergence rather than the artificially separated parts of it prior to their engagement. And to keep redefining it as it takes place – focusing on how such configurations take place and come to matter, how they are 'being produced through a series of entangled relational possibilities' (Malone 2015b, 8; also Rotas 2015).

Even for a posthumanistically oriented thinker shops seem, at first, an antithesis of 'nature'. Physically completely human built, with some natural (i.e. compostable) materials but mostly metals, plastics, cement, gravel and stones excavated and transported from quarries; flattened-out asphalt covered yards with parking spaces for cars and insides of shops filled with over-packaged produce, most of which you don't even need but are made to crave by global marketing schemes. However, for children in the North of Finland indoor shop spaces are a necessity directly related to and relating them with nature. Their surroundings are covered in ice and snow for half the year. Their concrete, daily relations to/with their natural surroundings are not static but change and evolve with seasons and between years – each winter is a bit different. The temperatures and slippery surfaces alone make children's movements change, the distances they walk become shorter, the time spent outside decreases and layers of puffy clothing become between their bodies and the outside world. Nature, as it were, drives children indoors during winters. Unless there is a shop, a library, a warm public space the children can pop in to warm up every now and then.

Just as the child–within–nature configuration of the 'shitgull' covered a wider and more complex assemblage than one child and one bird, the child–within–nature configuration of the 'shop' stretches to cover the means and movements with which children coexist within a particular setting. When temperatures fall, snow and ice set in and the sun doesn't rise for long thicker layers of clothes, indoor spaces and artificial lighting become part of child–within–nature configurations.

It's 6 pm in the evening. The children participating in my study have been running along bike paths in −26 °C temperature for half an hour. Everyone's freezing but nobody wants to go home. My car is not big enough to get us all inside it to warm up. They ask me if we could go to the nearby grocery shop. I am reluctant as I am supposed to be studying their engagements with their surroundings outside.

But I can't argue with freezing toes and so we make our way into the shop. I turn my recorder off and phase out my researcher mode, thinking we're taking a pause. After a good five minutes we're ready to go out again. As the children outrun me to climb the high snow pile at the rear of the parking lot I realise that there was no pause. That at this moment, with this temperature and these conditions, the child-within-nature configuration included the shop – the electric heating, neon lights, plastic wrapped candybars and Chiquita bananas – it had to. Importantly enough, another moment, another combination of children and seasons and surroundings would have emerged differently. Here we can also think of another formulation of posthuman politics: the shop, often understood in terms of consumerist desires, appeared as becoming part of the desire to spend time in contact with snow and cold air and peers in the evening. The shop thus had certain intensities in the assemblage, providing a place to warm up, but potentially also to enjoy different materials through senses, an escape, or an experimentation. These were mixed with the intensities, or affectual capacities, working between children, snow, and cold air. On surface level an encounter with a flying bird is easy to categorise as a desirable encounter with nature. Time spent indoors in a shop is as easy to dismiss from that category. The two events portrayed duly complicate and challenge such categorisations.

A posthumanist focus on the 'dance of agency' (Pickering 2005, 35) in both of the described events is a focus on the evolving of a particular process in time, a particular child–within–nature configuration rather than a focus on atemporal regularities. The political implications of this 'third way' (if the 'first' is a focus on things/material reality and the 'second' on meanings/social constructions both of which can yield atemporal regularities) are lodged in the ability of this approach to create knowledge of how phenomena emerge. How do children's relations to their urban environments emerge and evolve, today, *given all the complex socio-material-political conditions*. The politics of a posthumanist child–within–nature reconfiguration lies in the very possibility of reconfiguring subjectivities, producing entirely new subjectivities and 'ways of doing that have yet to be coded into heteronormative and/or disciplinary models' (Snaza et al. 2014, 48; Braidotti 2013). Posthuman politics is thus more of an interruption, an escape, an experimentation, rather than that of an ordering and division, because the polis is considered a more-than-human affair, working through situational intensities.

Conclusion

Statements such as 'urban children have lost their connection to nature' imply an anthropocentric view of humans as categorically separate from nature. The responses to such statements, albeit well-meaning, often reproduce the categorical divide with initiatives of simply reinserting the 'child back into nature' (Malone 2015b). Such an approach does little to further our understanding of child–nature relations as continuing transformation of both 'child' ('human') and 'nature' – and realising that essentially there is nothing to insert a child back into that would have preceded their engagement. And so, we argue that urban children can be thought of as particularly advantaged as subjects and agents of environmental education. They might be lacking with direct contact to what minority world children's societies call 'nature' but they are in the middle of complex political, cultural, and societal practices of engaged and mutually emergent coexistence with/as nature: sewage systems, dense populations on scarce land, floods, high-rises, decay, contrived parks, shops, lights, pigeons, rats, waste disposal systems, street dogs and dog parks. Their everyday lives are what Haraway (2016) and Tsing (2015) write about as staying with the trouble or living with the ruins – not as cynical quietism but true response-ability. This is what environmental educators can cultivate in urban settings.

When focused on measuring the connectedness of two separate entities, 'child' and 'nature', the research designs and pedagogical advances of environmental education are programs that seek to connect the two, often without questioning the very entities they are seeking to connect. For example, in a recent evaluation study of a *Get to Know Program* (Bruni et al. 2017), the main method for measuring connectedness was a computer game called the *Flexi Twins* game in which participants sorted words into two category sets: 'nature' or 'built', and 'me' or 'other'. The words selected for 'nature' included tree, mountain, butterfly, and flower. Words for 'built' were boat, car, chair, and truck. The underlying premises of the study are that we know objectively and universally – perhaps due to developmental theories – what 'a child' is, and also what is nature, and what is not nature. And what is excluded from 'nature' are often human-made artefacts, even those invented for the purpose of being with and connecting with natural elements, like boats to water. The binaries are evident, and the connections universally measured, regardless of whether a child has ever seen a mountain or not. The level of implications offered is often that of determining 'what frequency and duration [of direct contact] is needed' to enhance connectedness (Ernst and Theimer 2011, 593). And still, the conundrum remains that direct contact with natural surroundings (however defined) does not necessarily predict pro-environmental behaviour – the objective of environmental education – very well at all (Kollmuss and Agyeman 2002).[3]

To understand what makes us care and act responsibly is a challenge far more complex than the frequency and duration of direct contact with mountains and flowers. We argue that it is grounded in mutual emergence rather than fixed categories. Our conceptions of children and of nature are relative to each other and in constant flux, operating within complex socio-cultural-historical-ecological contexts. An ongoing discussion within education and childhood studies is challenging the longstanding reductionist tendencies of fixed categorising children, childhoods and ways of being a child (Lenz-Taguchi 2010; MacLure 2013). What is suggested instead of reductive categories is attending to the 'biosocial hybridity' of childhoods (Prout 2005), the 'excluded middle' between nature and nurture (Prout 2011) or in the words of Taylor, Pacini-Ketchabaw, and Blaise (2013, 81) attending 'to the challenges of growing up in an increasingly complex, mixed-up, boundary blurring, heterogeneous, interdependent and ethically confronting world'. To this end, Andrew Pickering (2005, 35) suggests that changing our 'unit of analysis' to a posthumanist one exposes a new field of environmental studies. Whether a new field or just a collection of converging approaches (see Snaza et al. 2014) it is nevertheless one that could not have been arrived at by following either of the traditional approaches (natural sciences or social sciences) individually or in combination.

The two child-within-nature events portrayed in this paper have been configurations of mutual emergence of matter and discourse, of subjects and objects, animate, inanimate, human, non-/more-than-human. That the relation of child and a seagull can be based on death is an indication of a society in which some of its members can be shot and others detained from cursing in the name of becoming human (less animal). Understanding 'shitgull' as an event, as the unit of analysis, makes it possible to critically review the complex conditions in which similar events arise rather than trying to find fault and inject cure to what are thought of as a priori participants: 'child' and 'bird'/'nature'. That an endangered Lesser black-backed gull is caught in a net and starves to death above a city owned human waste compost, and that a child (and his father) consider his death appropriate is ultimately not an indication of moral decay on the part of the individual humans, or of misguided feeding habits of the seagull. That is, the undesired event is not fixed by educating the humans or rerouting the seagull. This would be a temporary fix. Because the society that gave rise to the event – the conditions that made it possible – are still in place.

Education in general and environmental education in particular need ways to anticipate and evaluate the constantly emerging child–within–nature configurations so that some can be deemed more desirable than others. And more importantly: this evaluation can be done with the children, in their environments, as an integral part of environmental education itself. Rather than focusing effort on only educating the human individual to become more aware and knowledgeable of the (separate) nature, rather than only 'returning' children to the woods, environmental education research and practice could and should intensely focus on the everyday materialisations of complex historical, societal, political and cultural conditions that give rise to environmental phenomena, human attitudes and relations included. We need to ask, together with children (and animals and nonhuman surroundings) – How did we get here? What has happened that this is possible? Is this what is good-enough for all?

Notes

1. A thorough theoretical literature review of feminist and/or posthumanist scholars in various disciplines theorising human–animal (or human–nature) relations can be found in a recent issue of this journal (Lloro-Bidart 2015).
2. MacLure's method was crafted as part of her project to create research practices that could be labelled 'baroque' – resisting clarity, mastery, linearity, and coherence and honouring the details of educational events rather than rising above them. This was her response to the aggressive regulation of intellectual acts disparaged by the 'evidence-based practice movement'. Baroque methods or practices would resist the closure-seeking tendencies by remaining irritating and interuptive, in order to remain open to new questions.
3. This is not to say that direct contact with diverse environments would be of no value per se, or that no causal relation to a person's pro-environmental behaviour would exist.

Disclosure statement

No potential conflict of interest was reported by the authors.

References

Agamben, G. 2004. *The Open: Man and Animal.* Standford, CA: Standford University Press.
Alach, H. 2003. "Furry, Feathered, Finned or Feral: Pets in Early Childhood Settings." *The First Years: Na Tau Tuatahi. New Zealand Journal of Infant and Toddler Education* 5 (1): 25–27.
Atkinson, K. 2015. "Wasps-Bees-Mushrooms-Children: Reimagining Multispecies Relations in Early Childhood Pedagogies." *Journal of Childhood Studies* 40 (2): 67–79.
Barad, K. 2007. *Meeting the Universe Halfway: Quantum Physics and the Entanglement of Matter and Meaning.* Durham, NC: Duke University Press.
Bergson, H. (1903) 2007. *An Introduction to Metaphysics.* Edited by J. Mullarkey & M. Kolkman. New York, NY: Palgrave McMillan.
Braidotti, R. 2013. *The Posthuman.* Cambridge: Polity Press.
Brinkmann, S. 2014. "Doing without data." *Qualitative Inquiry* 20 (6): 720–725.
Bruni, C. M., P. L. Winter, P. W. Schultz, A. M. Omoto, and J. J. Tabanico. 2017. "Getting to Know Nature: Evaluating the Effects of the Get to Know Program on Children's Connectedness with Nature." *Environmental Education Research* 23 (1): 43–62.
Clarke, D., and J. Mcphie. 2014. "Becoming Animate in Education: Immanent Materiality and Outdoor Learning for Sustainability." *Journal of Adventure Education and Outdoor Learning* 14 (3): 198–216.
Cole, M., and K. Stewart. 2014. *Our Children and Other Animals: The Cultural Construction of Human–Animal Relations in Childhood.* Farnham: Ashgate.
DeMello, M. 2012. *Human–Animal Studies: A Bibliography.* Brooklyn, NY: Lantern Books.
Denzin, N. K. 2001. *Interpretive interactionism.* Thousand Oaks, CA: SAGE.
Despret, V. 2004. "The Body We Care for: Figures of Anthropo-Zoo-Genesis." *Body & Society* 10 (2–3): 111–134.
Despret, V. 2016. *What Would Animals Say If We Asked the Right Questions?.* Minneapolis, MN: Minnesota University Press.
Ernst, J., and S. Theimer. 2011. "Evaluating the Effects of Environmental Education Programming on Connectedness to Nature." *Environmental Education Research* 17 (5): 577–598.

Fenwick, T., R. Edwards, and P. Sawchuck. 2011. *Emerging Approaches to Educationalresearch: Tracing the Sociomaterial*. New York, NY: Routledge.

Friesen, L. 2010. "Exploring Animal-Assisted Programs with Children in School and Therapeutic Contexts." *Early Childhood Education Journal* 37: 261–267.

Gane, N. 2006. "Posthuman." *Theory, Culture & Society* 23 (2–3): 431–434.

Granjou, C. 2016. *Environmental Changes*. Elsevier: The Futures of Nature.

Halberstam, J., and I. Livingstone. 1995. *Posthuman Bodies*. Bloomington: Indiana University Press.

Haraway, D. 2003. *The Companion Species Manifesto. Dogs, People and Significant Otherness*. Chicago, IL: Prickly Paradigm Press.

Haraway, D. 2008. *When Species Meet*. Minneapolis, MN: University of Minnesota Press.

Haraway, D. 2016. *Staying with the Trouble: Making Kin in the Chthulucene*. London: Duke University Press.

Kahn, P. H. 2002. "Children's Affiliations with Nature: Structure, Development and the Problem of Environmental Developmental Amnesia." In *Children and Nature: Psychological, Sociological and Evolutionary Investigations* p. 93–116, edited by P. H. Kahn and S. R. Kellert. Boston, MA: MIT Press.

Kollmuss, A., and J. Agyeman. 2002. "Mind the Gap: Why Do People Act Environmentally and What Are the Barriers to Pro-environmental Behavior?" *Environmental Education Research* 8 (3): 239–260.

Law, J. 2004. *After Method: Mess in Social Scientific Research*. London: Routledge.

Lenz-Taguchi, H. 2010. *Going beyond the Theory/Practice Divide in Early Childhood Education: Introducing an Intra-Active Pedagogy*. London: Routledge.

Lewis, M. T. 2016. "Transcending the Student Skin Bag: The Educational Implications of Monsters, Animals and Machines." In *The Educational Significance of Human and Non-human Animal Interactions: Blurring the Species Line* p. 51–67, edited by S. Rice and A. G. Rud. Basingstoke: Palgrave MacMillan.

Liefländer, A. K., G. Fröhlich, F. X. Bogner, and P. W. Schultz. 2013. "Promoting Connectedness with Nature through Environmental Education." *Environmental Education Research* 19 (3): 370–384.

Livingston, J., and J. Puar. 2011. "Interspecies." *Social Text* 29 (1): 3–14. doi:10.1215/01642472-1210237

Lloro-Bidart, T. 2015. "Neoliberal and Disciplinary Environmentality and 'Sustainable Seafood' Consumption: Storying Environmentally Responsible Action." *Environmental Education Research*, 1–8. https://protect-us.mimecast.com/s/vleRBAsNvRJ9sa?domain=dx.doi.org. doi:10.1080/13504622.2015.1105198.

Lloro-Bidart, T. 2016. "A Feminist Posthumanist Political Ecology of Education for Theorizing Human–Animal Relations/Relationships." *Environmental Education Research*. 20(1): 111–130. doi:10.1080/13504622.2015.1135419.

Lorimer, J. 2012. "Multinatural Geographies for the Anthropocene." *Progress in Human Geography* 36 (5): 593–612.

Louv, R. 2005. *The Last Child in the Woods: Saving Our Children from Nature-Deficit Disorder*. Chapel Hill, NC: Algonquin Books.

Lupinacci, J., and A. Happel-Parkins. 2016. "(Un)learning Anthropocentrism: An EcoJustice Framework for Teaching to Resist Human-Supremacy in Schools." In *The Educational Significance of Human and Non-human Animal Interactions: Blurring the Species Line*, edited by S. Rice and A. G. Rud, 13–30. Basingstoke: Palgrave MacMillan.

MacLure, M. 2010. "The Offence of Theory." *Journal of Education Policy* 25 (2): 277–286.

MacLure, M. 2013. "Classification or Wonder? Coding as an Analytic Practice in Qualitative Research." In *Deleuze and Research Methodologies*, edited by R. Coleman and J. Ringrose, 164–183. Edinburgh: Edinburgh University Press.

Malone, K. 2015a. "Theorizing a Child–Dog Encounter in the Slums of La Paz Using Post-humanistic Approaches in Order to Disrupt Universalisms in Current 'child in nature' Debates." *Children's Geographies* 14 (4): 390–407.

Malone, K. 2015b. "Posthumanist Approaches to Theorising Children's Human–Nature Relations." In *Space, Place and Environment*, edited by K. Nairn, P. Kraftl, and T. Skelton, Volume 3 of *Geographies of Children and Young People*, edited by T. Skelton, 1–22. Springer Reference.

Mason, J., and B. Tipper. 2008. "Being Related: How Children Define and Create Kinship." *Childhood* 15 (4): 441–460.

McCardle, P., S. McCune, J. A. Griffin, and V. Maholmes. 2011. *How Animals Affect Us: Examining the Influence of Human–Animal Interaction on Child Development and Human Health*. Washington, DC: American Psychological Association.

McKay, R. 2005. "'Identifying with the animals': Language, Subjectivity and the Animal Politics of Margaret Atwood's Surfacing." In *Figuring Animals: Essays on Animal Images in Art, Literature, Philosophy and Popular Culture*, edited by M. S. Pollock and C. Rainwater, 207–227. New York, NY: Palgrave McMillan.

Melson, G. F. 2005. *Why the Wild Things Are: Animals in the Lives of Children*. Cambridge, MA: Harvard University Press.

Myers, O. E. 2013. "Children, Animals, and Social Neuroscience: Empathy, Conservation Education, and Activism." In *Ignoring Nature No More: The Case for Compassionate Conservation*, edited by M. Bekoff, 271–286. Chicago, IL: University of Chicago Press.

Nelson, N. 2016. Vibrant Matter(S): Figuring Death within Early Childhood Practices. Common Worlding with Children. A Common Wolrds Research Collective Blog. March 20, 2016. https://commonworldingwithchildren.com/2016/03/20/vibrant-matters-figuring-death-within-early-childhood-practices/.

Nelson, N., E. Coon, and A. Chadwick. 2015. "Engaging with the Messiness of Place in Early Childhood Education and Art Therapy: Exploring Animal Relations, Traditional Hide, and Drum." *Journal of Childhood Studies* 40 (2): 42–55.

Niccolini, A. D., and M. Pindyck. 2015. "Classroom Acts: New Materialisms and Haptic Encounters in an Urban Classroom." *Reconceptualising Educational Research Methodology* 6 (2): 1–23.

Oliver, K. 2009. *Animal Lessons: How They Teach Us to Be Human*. New York, NY: Columbia University Press.

Pacini-Ketchabaw, V., and F. Nxumalo. 2015. "Unruly Raccoons and Troubled Educators: Nature/Culture Divides in a Childcare Centre." *Environmental Humanities* 7: 151–168.

Pedersen, H. 2010. "Is 'the posthuman' Educable? On the Convergence of Educational Philosophy, Animal Studies and Posthumanist Theory." *Discourse: Studies in the Cultural Politics of Education.* 31 (2): 237–250.

Pedersen, H. 2015. "Parasitic Pedagogies and Materialities of Affect in Veterinary Education." *Emotion, Space and Society* 14: 50–56.

Pepperell, R. 2003. *The Posthuman Condition.* Bristol: Intellect.

Pickering, A. 1995. *The Mangle of Practice: Time, Agency, and Science.* Chicago, IL: University of Chicago Press.

Pickering, A. 2005. "Asian Eels and Global Warming: A Posthumanist Perspective on Society and the Environment." *Ethics & The Environment* 10 (2): 29–43.

Pink, S. 2009. *Doing Sensory Ethnography.* London: Sage.

Prokop, P., and S. D. Tunnicliffe. 2010. "Effects of Having Pets at Home on Children's Attitudes Toward Popular and Unpopular Animals." *Anthrozoös* 23 (1): 21–35.

Prout, A. 2005. *The Future of Childhood: Towards the Interdisciplinary Study of Children.* London: Routledge Falmer.

Prout, A. 2011. "Taking a Step Away from Modernity: Reconsidering the New Sociology of Childhood." *Global Studies of Childhood* 1 (1): 4–14.

Rautio, P. 2014. "Mingling and Imitating in Producing Spaces for Knowing and Being: Insights from a Finnish Study of Child–Matter Intra-Action." *Childhood* 21 (4): 461–474.

Rice, S., and A. G. Rud. 2016. *The Educational Significance of Human and Non-human Animal Interactions: Blurring the Species Line.* Basingstoke: Palgrave MacMillan.

Rocheleau, D. 2015. "A Situated View of Feminist Political Ecology from My Networks, Roots, and Territories." In *Practising Feminist Political Ecologies: Moving beyond the 'Green Economy'* p. 13–40, edited by W. Harcourt and I. L. Nelson. London: Zed Books.

Rotas, N. 2015. "Ecologies of Praxis: Teaching and Learning against the Obvious." In *Posthumanism and Educational Research*, edited by N. Snaza and J. Weaver, 91–103. London: Routledge.

Rowe, B. 2016. "Challenging Anthropocentrism in Education: Posthumanist Intersectionality and Eating Animals as Gastro-aesthetic Pedagogy." In *The Educational Significance of Human and Non-human Animal Interactions: Blurring the Species Line*, edited by S. Rice and A. G. Rud, 31–49. Basingstoke: Palgrave MacMillan.

Russell, C. L. 2005. "'Whoever Does Not Write is Written': The Role of 'Nature' in Postpost Approaches to Environmental Education Research." *Environmental Education Research* 11 (4): 433–443.

Russell, C. L., T. Sarick, and J. Kenelly. 2002. "Queering Environmental Education." *Canadian Journal of Environmental Education* 7 (1): 54–66.

Schultz, P. W. 2002. "Inclusion with Nature: Understanding the Psychology of Human–Nature Interactions." In *The Psychology of Sustainable Development*, edited by P. Schmuck and P. W. Schultz, 61–78. Boston, MA: Kluwer Academic Publishers.

Snaza, N., P. Appelbaum, S. Bayne, D. Carlson, M. Morris, N. Rotas, J. Sandlin, J. Wallin, and J. Weaver. 2014. "Toward a Posthumanist Education." *Journal of Curriculum Theorizing* 30 (2): 39–55.

Snaza, J., and J. Weaver. 2014. *Posthumanism and Educational Research.* London: Routledge.

Sørensen, E. 2009. *The Materiality of Learning: Technology and Knowledge in Educational Practice.* Cambridge: Cambridge University Press.

Taylor, A., M. Blaise, and M. Giugni. 2013. "Haraway's 'bag lady story-telling': Relocating Childhood and Learning within a 'post-human landscape.'" *Discourse: Studies in the Cultural Politics of Education* 34 (1): 48–62.

Taylor, A., and V. Pacini-Ketchabaw. 2015. "Learning with Children, Ants, and Worms in the Anthropocene: Towards a Common World Pedagogy of Multispecies Vulnerability." *Pedagogy, Culture and Society* 23 (4): 507–529.

Taylor, A., and V. Pacini-Ketchabaw. 2016. "Kids, Raccoons, and Roos: Awkward Encounters and Mixed Affects." *Children's Geographies* 15 (2): 131–145.

Taylor, A., V. Pacini-Ketchabaw, and M. Blaise. 2013. "Editorial. Researching Naturecultures of Postcolonial Childhoods." *Global Studies of Childhood* 3 (4): 350–354.

Thompson, K. L., and E. Gullone. 2003. "Promotion of Empathy and Prosocial Behaviour in Children through Humane Education." *Australian Psychologist* 38 (3): 175–182.

Tipper, B. 2011. "'A dog who I know quite well' Everyday Relationships between Children and Animals." *Childen's Geographies* 9 (2): 145–165.

Tsing, A. 2015. *The Mushroom at the End of the World: On the Possibility of Life in Capitalist Ruins.* Princeton, NJ: Princeton University Press.

Vidović, V. V., V. V. Štetić, and D. Bratko. 1999. "Pet Ownership, Type of Pet and Socio-emotional Development of School Children." *Anthrozoös* 12 (4): 211–217.

Watson, G. 2006. "Wild Becomings: How the Everyday Experience of Common Wild Animals at Summer Camp Acts as an Entrance to the More-than-Human World." *Canadian Journal of Environmental Education* 11: 127–142.

Weaver, J. 2010. *Educating the Posthuman: Biosciences, Fiction, and Curriculum Studies.* Rotterdam, NL: Sense Publishers.

Weil, K. 2010. "A Report on the Animal Turn." *Differences – A Journal of Feminist Cultural Studies* 21 (2): 1–23.

Thinking with broken glass: making pedagogical spaces of enchantment in the city

Noora Pyyry

ABSTRACT

In this paper, I explore thinking that happens in children's meaningful engagement with the city. To open up my argument, I discuss two events during which children are caught up in *intra -active* play with things and spaces. I argue that this mode of being joyfully engaged with one's surroundings is key to what Jane Bennett (2001) calls *enchantment*. This experience can be described as a sudden moment of wonder-at-the-world: it is an inspiring event, of being moved by something. It is a disruption that can open up new reflection. Because enchantment is highly affectual, it deepens one's engagement with the world: it fosters *dwelling with*. By this, I refer to making a home for oneself in the world, with the world. I approach this engagement and thinking with an acknowledgement of the capacity of the material and non-human world to provoke effects in human bodies: things and spaces thus take part in meaningful everyday encounters that make *dwelling with* possible. This more-than-human understanding allows for alternative ways of conceptualizing learning. Clean-cut categorizations such as 'learner', 'urban', 'nature', and so on become problematic, and learning is re-conceptualized as an ongoing, non-linear and rhizomatic event in which knowing and being are always tied together. While playing, children are open to the unexpected: they are dwelling with the city and take part in creating new *pedagogical spaces of enchantment*.

Being enchanted is passion about dwelling with the world

I will start my discussion with a mundane event from San Francisco, California, where I was doing participatory research with tween girls on their urban geographies of hanging out (e.g. Pyyry 2015; Pyyry 2016b). As social science researchers know from experience, academic work can never really be separated into phases of fieldwork and analysis, fieldwork or no fieldwork. In fact, it is often difficult to divide the work into separate projects or detach it from other spheres of life. So, although I was conducting a participatory study with a group of tween girls, the research impacted all of my life. The vignette, which I will soon turn to, was an event that captured my attention, since I was working with Tim Ingold's (2000) concept of *dwelling with* in order to explore human involvement with the urban landscape. Something happened when I was walking home with my children one day. Looking back, it was a moment of the world speaking back to theory. Consequently, the vignette opens up the argument of this paper particularly well: it suggests that thinking emerges in unforeseen encounters. In research, empirics and theoretical analysis always go together, and they get entangled with everything else that is going on. It is in this emergent, rhizomatic *relational field* that researchers build their understandings (Hultman and Lenz Taguchi 2010). We know through engaging, through dwelling with the world.

Ergo, with the vignette, I want to illustrate what I think is important in conceptualizing dwelling with, and the openings this involvement can create for *enchantment* (Bennett 2001). These two inter-linked concepts are essential to how I attend to thinking and learning, to being playful and engaged with the world. In the event that captured my attention in San Francisco, my son noticed broken black glass on the street and stopped to play with it (Figure 1). He was completely caught up in the play, he did not really hear or see anything else that was happening around him. He was feeling the glass in his hands, collecting the pieces and examining them. In this moment, I argue that he was affectually engaged with the street and the glass so that he was becoming something else. He was hence not a single, coherent subject engaging with separate objects, rather I would say he was a coming-together of what was going on there and then. Here, 'the subject' and the 'object' fall on the same side of a shared movement, as Massumi (2011) puts it: they are linked in a joint-experience of moving-toward. This is an *event*: 'a rolling of subjective and objective components into a mutual participation, co-defining the same dynamic'. Thinking about the world through events carries with it an understanding that the subject and object are always connected. There is no such thing as a 'perceiving subject' outside of the world, rather the subject is created through events, again and again.

While playing with the pieces of glass – or while the glass was playing with him – my son was extremely happy and seemed to be quite in awe of the world, immersed in it. I argue that this mode of being joyfully and meaningfully engaged with spaces and things is key to the unusual experience of enchantment. This is 'dwelling with'.

In thinking of the transient moment of enchantment, the concept of 'dwelling with' has proved to be valuable (e.g. Pyyry 2016a, 2016b). Dwelling with, as I illustrated with the vignette, refers to being meaningfully engaged with one's material environment, and more broadly with the world. It is making a home for oneself in the world, with the world. Understood this way, dwelling is always connected to building: they are two sides of the same coin (Rose 2012). When 'building' is conceptualized broadly, it includes small passing acts, such as children's everyday spatial appropriations. Through these acts, together with the material and the non-human world, children produce alternative modes of involve-ment with the city and temporarily claim the world as their *own*. This clears space for enchantment.

The experience of enchantment is an uplifting delight about being alive. It is a feeling of wonder-at-the-world. Although often connected with joy, it can also be a state of trouble, even fear (Bennett 2001). The broken glass intrigued my son, but it could have also posed questions to him in the form of small pain. So, enchantment does not have to be a pleasant feeling of being connected with the world, the point is that enchantment has a strong affective force. In effect, enchantment is often a moment of simultaneous *immersion* and *disconnect* with the world: it's a sort of 'opening-onto' and 'distancing-from' the world (Wylie 2009). So, it is not only a fusion of self and the world, rather it is both presence and absence at the same time. This unspecific moment of mystery often goes by unnoticed, it is a short-lived juncture that is always already escaping. Enchantment is often out of the reach of verbal representation, but it is a powerful force that is *felt,* you are sort of struck by it when caught up in a moment. So, enchantment is not something we can necessarily pin-point or prove to have happened: it is non-representational.

Because enchantment is highly affectual, it deepens human engagement with the world, with places and things. Enchantment and dwelling, then, nurture each other. Enchantment attunes you to the world differently, it shakes your subjectivity. It is a moment of feeling the world running through you (see Massumi 2000). Hence, Bennett (2001) convincingly argues that enchantment is key to ethical being-in-the-world. So, here is where love comes to the picture: when you are deeply involved with something, you tend to care for it too. This understanding places value to spending time with ordinary, everyday things and spaces. Enchantment is then not so much connected with curiosity (obsession with the new), but with genuine wonder (Stone 2006). It emanates from dwelling with, which can be characterized as a sort of opening to the world. Involved activity strengthens the spatial relationship, and respect for the urban environment is built in this process.

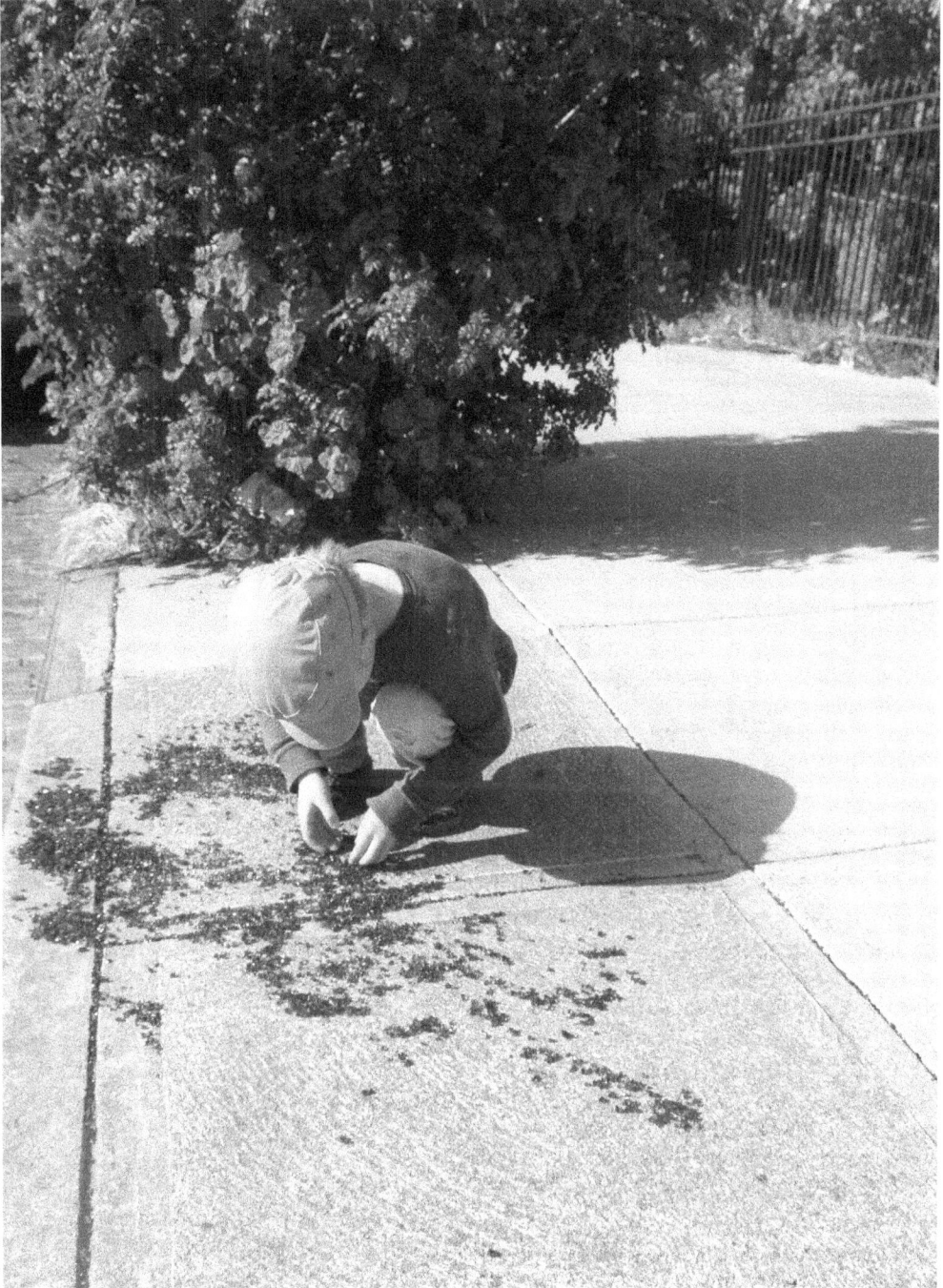

Figure 1. Learning with the city in San Francisco.

In the event of my son engaging with the glass, learning emerged from the joint-experience: the pieces of glass, the street, sounds, scents and everything that came together there and then took part in the event. There was then no individual 'learner', no 'subject' to think the world from the outside: thinking happened with the world. Karen Barad (2003, 803) talks about *intra-activity*, when she discusses the mingling of different elements that do not have clear boundaries. The concept is an important effort to go beyond thinking about the world through boundaries and dichotomies. A piece of glass can take part in intra-active play, from which new spaces can be created. A human body, as all bodies, can then be understood as a fluid, ongoing and relational nexus of flows, inseparable from the rest of the world: it is a coming-together of forces (Grosz 2005). Human agency emerges differently in different situations, since it is never thoroughly an individual affair. Therefore, knowing and learning are understood to take place *within* the world, with everything that comes together in an event (e.g. Fenwick, Edwards, and Sawchuk 2011; Taylor, Blaise, and Giugni 2013). Although I grant a special role to human intentionality in this paper's argument, my son's body and agency thus emerged differently in this situation than they would have done, for instance, at a playground. Subjectivity is produced in encounters with the world, and it is therefore important to pay attention to the kind of encounters our cities allow for.

'The bubble-wrap generation' is living and learning in tight time-spaces

'Time is life, and life resides in the human heart. The men in grey knew this better than anyone. Nobody knew the value of an hour or a minute, or even a single second, as well as they. [...] They had designs on people's time – long-term and well-laid plans of their own. What mattered most to them was that no one should become aware of their activities. They had surreptitiously installed themselves in the city.' (Ende 1973/1984)

This quote from Michael Ende's (1973/1984) famous novel, *Momo,* tells about grey men, who are stealing time from humans. The whole city in the story becomes dull, sterile and devoid of all things that are not considered useful. The grey men steal all the fun in the world: they steal the possibility for enchantment. Only Momo, a little girl with extraordinary abilities, can fight them. The child eventually rescues all the people and brings life back to the city.

The vignette of my son engaged in play with the glass can be approached from many points of view. One of these is the increased scheduling and monitoring of children's lives in the name of safety and productivity. It is understandable that not all parents would want their children to play with broken glass on the street. A playground is viewed as much more suited to a young child. Malone (2007) has called children of today 'the bubble-wrap generation' to bring attention to the highly controlled environments children live in, also in their free time. 'Security' is used to justify restrictive policies and practices in urban spaces, and the tightened notions of safety strongly affect children's lives in Western cities (e.g. Pyyry and Tani 2016). Children's days tend to be scheduled very early on for 'educational' and 'developmental' purposes and parents often end up driving them from one organized activity to another (see Karsten 2005; 'the backseat generation'). As a result, very little time is left for children's independent mobility and free, unplanned experimentation with everyday spaces. Children also quickly internalize shared notions of safety and become more hesitant to explore new spaces. Upsetting news of crimes against children, or horrible stories about accidents that are waiting to happen, further socialize them into the security talk.

When fear guides urban planning and parental decision-making, there is a risk that children are kept supervised at all times in spaces specifically designed for their use (e.g. playgrounds). Aitken (2001) remarks that children are actually spatially outlawed by society in these acceptable 'islands'. Even when playspaces are built for more extreme fun and entertainment, the experience is often designed by authoritarian capitalism – an interesting mix of control through surveillance and distraction through entertainment (see Thrift 2011). These spaces cultivate political passivity. And, when people are charged for entering, many children are placed as outsiders. Although playgrounds are often pleasant and convenient places for parents and children to gather in, they take part in the re-production of the pre-vailing urban order. Playground spaces can be effectively used for their designed purpose, but within the functional planning, improvisation often feels unwelcome and children therefore conform to the established norms of movement. If these norms and materialities are taken for granted, children may not have much space for playing with who or how they are. Franck and Stevens (2007) note that even when these types of *tight spaces* could be used differently, this may not be tolerated by the majority of people. Playgrounds may physically allow for many different kinds of activities, but they tend to be selective in terms of who is invited (according to age etc.) and which ways of being/doing are welcomed. To make these spaces *looser* would require being differently in them.

Not only do children spend their free time in tight spaces, but often from an early age, they also attend to some form of 'schooling'. Formal education is structured around time-spaces within which it is difficult for anything radically new to emerge. It is obviously artificial to make a division between 'formal' and 'informal' learning environments, since as everything else, these overlap and connect in many ways. Still, despite the skills based reforms that are taking place in many Western countries, the formal educational structures often serve very traditional notions of learning. The aim of this article is not to make a statement about what formal geography or environmental education should look like (for this, see e.g. Lambert, Solem, and Tani 2015; Wadley 2008), but to sketch what learning is in a more-than-human urban context and think of how it comes forth in the world. Two aspects are particularly important. The first is the dominant emphasis on learning as something that an individual human body does. He/she acquires information about the world (that exists 'out there'). This implies that humans are separate from the world. This separation brings with it a belief that learning can be measured by testing individual performance. Coupled with this is the predominance of relatively fixed and pre-given meanings and normativities of learning as deliberative, useful, discursive, and something that occurs in particular times and places, in a linear manner. Even when learning is understood in a more flexible way, many of the defining structures (e.g. standardized testing) remain. As the vignette in the beginning suggested, there is more to learning than gathering information *about* the world. Placing the human being *in* the world means overcoming the subject-object dichotomy that has dominated Western thought for too long. Re-conceptualizing learning in a more-than-human, practice-oriented frame makes it possible to take it as an orientation toward futures, not as an act of distancing (the world from the self).

This conceptualization gives value to children's everyday practices. It also places importance on children's right to their city. This relates to what Lefebvre (1947/2014) and the avant-garde group Situationists International strived for with their ideal of 'unitary urbanism'. The Situationists saw functionalist urban planning as a threat for spontaneity and wonder, and aimed to shake the urban order with imagination and play (see Knabb 2006). They linked action and understanding in their practices of studying the city, just as children do when playing with interesting materials. Following their thought, children – and all people – should have time and space to engage with urban spaces, to be affected by them. Spatial-embodied transformation comes from everyday practice and involved activity fosters care for the urban environment. The right of access and care for one's city thus go together.

When engaging with broken glass, the surface of the street or other interesting materials, children learn to respond to their environment wisely. Learning happens while being affectually engaged with spaces and things, in intra-active play. This is what Ingold (2000) calls enskilment: learning in everyday practice. Enskilment goes together with place-making, a creative process of engagement with the world that cultivates spatial skills. Because movement and cognition are tied together, this practical knowledge cannot be taught outside the context of use. And, it cannot be measured by tests, such as the PISA (the OECD Programme for International Student Assessment). This is why this multisensory reflection about one's place in the world is not always recognized and valued at a time of security talk and pressures of productivity. It is as if Ende's horrific fantasy-world has come true.

Entangled with 'other' living material and creating new worlds

My second vignette is from the Oxford University grounds, England. Here (Figure 2), my two children climbed up a stone wall with the help of a tree conveniently growing close to the wall. From a post-human point of view, the children did not just decide to climb up – rather, the intriguing setup invited them to explore. Bennett (2010) refers to this as *thing power*: the material and non-human world has a capacity to affect human bodies. This idea entails an acknowledgement of the liveliness that is internal to materiality, i.e. also the stone wall. But, not only did the wall and the tree actively invite the children to climb, in the event, they all became something more, something new: the children's bodies, the tree, the wall, ideas of safety, gender or appropriate behavior in the university area, my fear, other people passing by, and much *more* mingled without clear boundaries. This 'more' includes also things absent

Figure 2. Learning with the city in Oxford, England.

at that moment. They all took part in intra-active play from which new worlds (and bodies) are created. This is new knowledge. It emerged relationally, from the mix of human intentionality and other forces of the world.

A relational conceptualization of the world makes rigid categorizations such as 'nature/urban' or 'nature/culture' impossible. The world is understood to be more complex than that. It emerges through events as an ongoing (be)coming-together of multiple things (matter, action, relations). The children-tree-wall combination was a rhizomatic, multidirectional unit of becoming through which agency, power and thinking actualized. Pauliina Rautio (2013) discusses children's internally rewarding everyday practices, such as carrying stones, to open up how we are continually constituted by other animate or non-animate entities. This means that humans are always nature and culture, and it is impossible to make the division. The children-tree-wall unit was not a stable one, it was an open-ended formation that was also intra-active with everything else taking place then. Effectivity and knowing emerged rhizomatically within the flows of energy in the climbing event. There is, of course, no way of proving that enchantment took place there and then, but the atmosphere of the event that 'vibrated in the air' suggested that something significant happened. There was a burst of joy that came from this joint participation that could also be felt behind the camera. 'Joy' here is understood as something more than just a human psychological state, it is a wave of uplifting ontological energy (Braidotti 2013). This affectual atmosphere and feeling 'points to the possibility that minor variations within and between bodies may generate affective turbulence producing in turn a sense that some kind of difference is in the making', as McCormack (2013, 33) puts it. The feeling is non-representational.

Non-representational geography or non-representational theory (NRT; see Anderson and Harrison 2010; McCormack 2013; Thrift 2008, 2011; Wylie 2009) provides theoretical tools for thinking about learning as a spatial-embodied transformation that is prompted by everyday encounters. This trans-formation emerges in events, and the multiple forces at work cannot be fully captured with verbal description. Attention needs to be paid to the performative and affectual, since representationalism separates the world into the ontologically disjointed realms of words and things (see Barad 2003). This separation pervades the dominant understanding of learning as an individual human project that can be represented with evidence (and thus measured with tests). Non-representational theorization calls for attention to things taking-place: it aims to work beyond representation, because it looks at action as a relational phenomenon. Involvement with the world is habitual rather than 'conscious'. Non-representational theorization thus aims to study the multiple spatial processes that are involved in an event, whether or not they cross the threshold of contemplative cognition (McCormack 2003). Emphasis is therefore placed on everyday affectual geographies that are often left unnoticed, but through which the world is experienced (e.g. Horton and Kraftl 2006). For children, an important everyday practice and a way of relating with the world is play.

Towards playful cities

Children's play has been studied from many perspectives and has been given numerous functional definitions over time, ranging from evolutionary to sociological interpretations (see Sutton-Smith 1997). Often, play has been connected with the development of a child (e.g. Mead 1934; Piaget 1962; Vygotsky 1967). Conceptualizing play with non-representational theory makes it possible to view it as a 'mode' of being in the world, rather than a form of behavior that can be observed from the outside (Thrift 2011). Play is openness towards the world and receptiveness to new ideas: it is a joyous way of engaging with the world. Looked at this way, play is not examined as a means to an (educational or developmental) end; instead it is valued as such. Play is complicated, undefinable and de-individualistic, it consists of many different forms of being (Rautio and Winston 2015). Most importantly, play is highly affectual, and it therefore holds the capacity to change things. When involved in meaningful playful activities, children are open to the world unfolding, they are open to the complexity of life and the unforeseen. They engage with things and spaces in a non-instrumental, goal-free way. This is why play fosters dwelling with: it is pleasantly purposeless (from an adult point of view) and nomadic. In play, children make the city their playground, and home, through improvisation and experiment.

This connects the discussion to spatial criticism and children's right to the city. Stevens (2007) talks about 'urban play' when he examines people's spontaneous creative acts and spatial appropriation in the city. Children's urban play and willingness to experiment with their surroundings is just what the Situationists called for when they were hoping for a revolution that would arise from everyday life.

Play carries with it power to open up space for enchantment and everyday politics, for re-thinking the city and for transformation of space. In play, children speak back to the adult world: they imagine new ways of engaging with it. Play thus entails potential for being *otherwise*. Children's urban play makes possible futures and ways of being visible to the rest of us, just as Momo, the little girl in Ende's story does with a help of her turtle-friend. Children invite us to fight the grey, time-stealing vampires of neoliberal everyday life by letting ourselves be caught up in a moment of enchantment (also Duhn 2016). Children dwell with the city, claim it as their own, and change it through small acts of 'building'.

The event of intra-active play at the Oxford University grounds re-worked the space, even if just momentarily. Through the playful encounter, the children challenged the routine ways of movement there. Spaces of learning with the city can be conceptualized as experiments with the world that is con-tinually in-the-making. This conceptualization draws attention to the disjointed and emergent nature of learning. By shifting the attention away from the individual knowing subject, learning is reconceptual-ised as a non-linear and multidirectional process, it is a complex coming-together of things taking place then and there (also Fors, Bäckström, and Pink 2013). In the event of climbing, action and reflection was provoked by the encounter with the non-human. The feel of different surfaces and other living material invited the children to intra-active play from which new *pedagogical spaces of enchantment* can emerge (Pyyry 2016a). Learning happened through dwelling with the city: it was spatial-embodied. This links to what I have elsewhere conceptualized as 'hanging out -knowing': an ongoing process of reflection about one's place in the world that takes place in everyday encounters (Pyyry 2016a). In a more-than-human frame, learning is an event of 'dwelling with – enchantment – reflection'. Many things take part in the event, and learning does not evolve in a temporal order. Rather, new knowledge emerges sur-prisingly. Involvement and being playful is important here, since joy inspires imagination and can take learning to new directions. Engaging with spaces and things makes it possible to see the world anew and to produce new means of association. It makes it possible to think that which cannot be verbalized.

Conceptualized this way, learning resembles life: it exceeds categories and simple definitions. Something always escapes. Learning via enchantment is fueled by the excess and complexity of the world. Enchantment is a moment when familiar things appear odd, even surreal, and it therefore opens up questions, about *how things stand* (Bennett 2001). Enchantment can never be predicted – it is always accidental – but by supporting children's spontaneous playfulness in cities, space can be cleared for the inspiring experience. This unspecific event of being moved by something provokes awareness of how the world is framed, of how it works – and most importantly, makes it possible to imagine *how it could be*.

Reflections

The argument of this paper is based on the recognition that knowing exceeds individual human beings: it is much more than a human issue. Rather than viewing the human being as a curious world-molding subject, the world poses questions to us as we go with it. Knowing takes place in shared movement, and our subjectivities are always connected to what is going on. New reflection on urban matters becomes possible through dwelling with the city and emerges in affectual moments of enchantment, which in turn, deepen a person's relationship with the world. In an event of 'dwelling with – enchant-ment – reflection', one is somehow differently attuned to being. This experience of not-understanding (wonder) takes time and space. Learning with the city is thus closely connected to questions of spatial politics, right to the city and care for the world. A more-than-human approach ties together action and reflection by placing the human subject firmly in the world, and hence supports critical environmental education practice, which faces tremendous challenges in today's world (e.g. Payne 2016).

When children have time and space to engage with the urban environment, they are often caught up in imaginative intra-active play with trees, stones or just broken glass on the street. In these moments, they seem to be immersed in the world, or at least deeply engaged with it: intrigued, puzzled and inspired. I argue that this receptive mode of being playfully engaged with everyday spaces and things makes it possible to be genuinely moved by the world. This highly affectual experience of enchant-ment is a disruption that can open up new reflection. Enchantment can be prompted by the strange and the novel, but most often it arises from encounters with the very ordinary. Something happens, sort of hits you: this is the force of the world. The moment opens up questioning, modest or profound, and produces new ways of being. New knowledge emerges. Moments of enchantment are always accidental, but by letting children wander and wonder – by letting them dwell with their cities – space is cleared for the inspiring experience. Then, children have a chance to question the routine, fleeting and taken-for-granted aspects of urban life. Through this, they become enskilled urban dwellers. When given time and space to do so, children experiment with what is intriguing, what is scary, what is dull – and, most importantly with how the world could be. Many things come together in these events of experimentation: materials, energy, human and non-human bodies, feelings of fear, excitement, joy and more, memories, ideas, hopes and wishes – also people and things absent affect the situations (this includes ideologies and hidden attitudes). Sensing and thinking emerge in the encounters, relationally. As Morgan (this issue) points out, it is crucial to recognize all different agents affecting children's urban geographies, so that these can be taken into account also in formal education and policy-making.

The vignettes I introduced in this paper may provoke discussion about safe and child-friendly urban spaces. I hope it will not emerge in the form of 'security talk', since within tight urban spaces and highly structured everyday lives, children have few chances to experiment with different ways of being-in-the-world. They have very few chances for creating their own pedagogical spaces of enchantment. By paying attention to the ways in which learning emerges relationally in complex and non-linear ways, it is possible to acknowledge and respect the countless ways in which children already know. And, it is possible to appreciate their desire to learn more. I therefore wish for adults to work hard on making space for children's alternative modes of experimentation with the city. Even, and especially, when this may not seem very clean or convenient. This call for attention is not meant as an attempt to integrate children's urban geographies into the educational system in any instrumental way, although it is possible to link school-work to children's creative means of exploring their city. Spatial-embodied knowing can be articulated at school through art, writing, drama or imaginative mapping of urban spaces that matter to children.

Child-friendly urban spaces are loose spaces: they allow for children to engage in intra-active play with spaces and things, and with other living material. Child-friendly spaces allow children to be *otherwise*. But, children also create these spaces. When engaged meaningfully with the material and the non-human world, children momentarily claim spaces as their own. They are linked with the city in an experience of joint-participation from which everyday politics can emerge. Through small and momentary practices, children question the rules of appropriate behavior and prevailing ideas of safety in the city, and through this, carve space for alternative ways of being. By dwelling with urban spaces children express what is important to them (see Rautio and Jokinen 2015). And, by probing the limits of everyday life, children make the city more interesting for adults, as well. They make space for diversity, spontaneity and playfulness. This is why the conceptualization of learning introduced in this paper is not only relevant to educators and parents, it is also an urgent call to urban planners to re-think what cities could be: places of openness and enchantment, provocation, (com)passion and generosity.

Disclosure statement

No potential conflict of interest was reported by the author.

Funding

This work has been funded by the Kone Foundation.

References

Aitken, S. C. 2001. *Geographies of Young People: The Morally Contested Spaces of Identity*. London: Routledge.

Anderson, B., and P. Harrison, eds. 2010. *Taking-Place: Non-Representational Theories and Geography*. Farnham: Ashgate.

Barad, K. 2003. "Posthumanist Performativity: Toward an Understanding of How Matter Comes to Matter." *Signs: Journal of Women in Culture and Society* 28 (3): 801–831.

Bennett, J. 2001. *The Enchantment of Modern Life*. Princeton, NJ: Princeton University Press.

Bennett, J. 2010. *Vibrant Matter: A Political Ecology of Things*. Durham: Duke University Press.

Braidotti, R. 2013. *The posthuman*. Cambridge: Polity Press.

Duhn, I. 2016. "Speculating on Childhood and Time, with Michael Ende's Momo (1973)." *Contemporary Issues in Early Childhood* 17 (4): 377–386.

Ende, M. 1973/1984. *Momo*. (Original work *Momo* published in 1973.) Translation by Brownjohn, J.M. New York: Penguin books.

Fenwick, T., R. Edwards, and P. Sawchuk. 2011. *Emerging Approaches to Educational Research: Tracing the Sociomaterial*. London: Routledge.

Fors, V., Å. Bäckström, and S. Pink. 2013. "Multisensory Emplaced Learning: Resituating Situated Learning in a Moving World." *Mind, Culture, and Activity* 20: 170–183.

Franck, K. A., and Q. Stevens. 2007. "Tying down Loose Space." In *Loose Space: Possibility and Diversity in Urban Life*, edited by K. A. Franck and Q. Stevens, 54–72. London: Routledge.

Grosz, E. 2005. *Time Travels*. Durham and London: Duke University Press.

Horton, J., and P. Kraftl. 2006. "Not Just Growing up, but Going on: Materials, Spacings, Bodies, Situations." *Children's Geographies* 4 (3): 259–276.

Hultman, K., and H. Lenz Taguchi. 2010. "Challenging Anthropocentric Analysis of Visual Data: A Relational Materialist Methodological Approach to Educational Research." *International Journal of Qualitative Studies in Education* 23 (5): 525–542.

Ingold, T. 2000. *The Perception of the Environment*. London: Routledge.

Karsten, L. 2005. "It All Used to Be Better? Different Generations on Continuity and Change in Urban Children's Daily Use of Space." *Children's Geographies* 3: 275–290.

Knabb, K., ed. 2006. *Situationist International: Anthology*. Berkeley, CA: Bureau of Public Secrets.

Lambert, D., M. Solem, and S. Tani. 2015. "Achieving Human Potential through Geography Education: A Capabilities Approach to Curriculum Making in Schools." *Annals of the Association of American Geographers* 105 (4): 723–735.

Lefebvre, H. 1947/2014. *Critique of Everyday Life*. The one-volume edition. (Original texts of *Critique de la vie quotidienne* published in 1947, 1961 and 1981). Preface by M. Trebitsch, translation by J. Moore, and G. Elliott. London: Verso.

Malone, K. 2007. "The Bubble-Wrap Generation: Children Growing up in Walled Gardens." *Environmental Education Research* 13 (4): 513–527.

Massumi, B. 2000. "The Ether or Your Anger: Towards a Pragmatics of the Useless." In *The Pragmatist Imagination: Thinking about Things in the Making*, edited by J. Ockman, 160–167. Princeton: Princeton Architectural Press.

Massumi, B. 2011. Conjunction, Disjunction, Gift. *Transversal. A Multilingual Webjournal*. Available from: http://eipcp.net/transversal/0811/massumi/en

McCormack, D. P. 2003. "An Event of Geographical Ethics in Spaces of Affect." *Transactions of the Institute of British Geographers* 28 (4): 488–507.

McCormack, D. P. 2013. *Refrains for Moving Bodies*. Durham and London: Duke University Press.

Mead, G. H. 1934. *Mind, Self, and Society*. Chicago, IL: University of Chicago Press.

Payne, P. G. 2016. "What Next? Post-Critical Materialisms in Environmental Education." *The Journal of Environmental Education* 47 (2): 169–178.

Piaget, J. 1962. *Play, Dreams, and Imitation in Childhood*. New York, NY: Norton.

Pyyry, N. 2015. "'Sensing With' Photography and 'Thinking With' Photographs in Research into Teenage Girls' Hanging out." *Children's Geographies* 13 (2): 149–163.

Pyyry, N. 2016a. "Learning with the City via Enchantment: Photo-Walks as Creative Encounters." *Discourse: Studies in the Cultural Politics of Education* 37 (1): 102–115.

Pyyry, N. 2016b. "Participation by Being: Teenage Girls' Hanging out at the Shopping Mall as 'Dwelling With' [the World]." *Emotion, Space and Society* 18: 9–16.

Pyyry, N., and S. Tani 2016. Young People's Play with Urban Public Space: Geographies of Hanging out. In *Play, Recreation, Health, and Wellbeing*, edited by B. Evans and J. Horton, *Vol. 9 of Geographies of Children and Young People*, edited by T. Skelton. Singapore: Springer.

Rautio, P. 2013. "Children Who Carry Stones in Their Pockets: On Autotelic Material Practices in Everyday Life." *Children's Geographies* 11 (4): 394–408.

Rautio, P., and P. Jokinen 2015. Children's Relations to the More-than-Human World beyond Developmental Views. In *Play, Recreation, Health, and Wellbeing*, edited by B. Evans and J. Horton, *Vol. 9 of Geographies of Children and Young People*, edited by T. Skelton. Singapore: Springer.

Rautio, P., and J. Winston. 2015. "Things and Children in Play – Improvisation with Language and Matter." *Discourse: Studies in the Cultural Politics of Education* 36 (1): 15–26.

Rose, M. 2012. "Dwelling as Marking and Claiming." *Environment and Planning D: Society and Space* 30 (5): 757–771.

Stevens, Q. 2007. *The Ludic City: Exploring the Potential of Public Spaces*. London: Routledge.

Stone, B. E. 2006. "Curiosity as the Thief of Wonder: An Essay on Heidegger's Critique of the Ordinary Conception of Time." *KronoScope* 6 (2): 204–229.

Sutton-Smith, B. 1997. *The Ambiguity of Play*. Cambridge, MA: Harvard University Press.

Taylor, A., M. Blaise, and M. Giugni. 2013. "Haraway's 'Bag Lady Story-Telling': Relocating Childhood and Learning within a 'Post-Human Landscape'." *Discourse: Studies in the Cultural Politics of Education* 34 (1): 48–62.

Thrift, N. 2008. *Non-Representational Theory: Space/Politics/Affect*. London: Routledge.

Thrift, N. 2011. "Lifeworld Inc – And What to Do about It." *Environment and Planning D: Society and Space* 29: 5–26.

Vygotsky, L. S. 1967. "Play and Its Role in the Mental Development of the Child." *Journal of Russian & East European Psychology* 5 (3): 6–18.

Wadley, D. A. 2008. "The Garden of Peace." *Annals of the Association of American Geographers* 98 (3): 650–685.

Wylie, J. W. 2009. "Landscape, Absence and the Geographies of Love." *Transactions of the Institute of British Geographers* 34 (3): 275–289.

'I saw a magical garden with flowers that people could not damage!': children's visions of nature and of learning about nature in and out of school

Clementina Rios and Isabel Menezes (iD)

ABSTRACT

This paper involves groups of children (aged 5–10) in discussing *what* nature is in their urban communities and *how* they learn about it. Children attend four urban and semi-urban Portuguese schools with different environmental pedagogies: Waldorf, forest school and eco-school. Previous studies of children's conceptions of nature have mainly addressed environmental understanding as an individual dimension, even if acknowledging the situated nature of children's knowledge and experience. In this study we draw on previous research, using focus groups as participatory methods that allow children to interact with their peers while expressing their visions and feelings about a topic. Group discussions show that children have a strong emotional connection with nature that generates a strongly protective disposition. Daily experiences in schools, families, and local communities but also the media reinforce this concern, and make children aware of a series of environmental problems, for which they either refer to existing rules or imagine creative solutions. On the whole, this research shows that children have a say in these matters and should therefore be involved in environmental debates and action – but also that a political ecology perspective seems to be absent from their school learning experiences.

Introduction

Children in urban environments are frequently kept apart from nature and isolated from the dangers of the outside world: traffic, violence, strangers, rain and cold. There is a decline in outdoor games, in- and out-of-school, and urban spaces rarely have green areas where children can freely experience direct contact with nature (Malone 2007; Wilson 1996). Even if fallacious, a vision of the 'outside world' as potentially dangerous is frequent, and keeping children inside is seen as protective and risk-free (Adams and Savahl 2015; Duhn 2012). Yet, childhood environmental experiences are central for their knowledge of the local community with 'very important aspects of learning including psychological, social, cultural, physical and environmental' (Malone 2007, 523). While scarce, some research with children shows that their representations of the community include both the built and the natural environment (Adams and Savahl 2015; Dockett, Kearney, and Perry 2012; Machemer, Bruch, and Kuipers 2008; Pooley et al. 2002), while research with adolescents confirms that sense of community is an important predictor of social and civic engagement and participation, and therefore positively related with youth citizenship

(Albanesi, Cicognani, and Zani 2007; Ferreira, Coimbra, and Menezes 2012). We can therefore wonder if growing up in a world where the 'green' – the woods, the forests, the 'wild' nature – is viewed as distant, exotic and even dangerous (Adams and Savahl 2015) has implications for the development of a sense of belonging and the capacity to emotionally connect with the larger global community. This is of particular significance if we consider the challenges of environmental issues that are at the core of citizenship today – leading Latour to call for a political ecology that will 'welcome nonhumans into politics' (2004, 226). While research shows that 'children have significant concerns for their local environment and a sense of injustice that their experience is not systematically considered' (Hacking, Scott, and Barratt 2007, 532), it is difficult to predict if this engagement will persist in the face of a growing detachment and isolation from nature.

Education for environmental sustainability

The emphasis on sustainable development is, 'a key issue for the educational system as a whole' (Hedefalk, Almqvist, and Östman 2015, 975), including the years of pre-school and elementary school, even if not always contemplating the need to articulate environmental issues with the political and civic dimensions of children's lives. In their review, Hedefalk, Almqvist, and Östman (2015) conclude that teachers see this as an opportunity to promote children's knowledge, behaviours and critical thinking, while favouring participatory and practical approaches that involve direct contact with nature and engagement with real-life problems; other promising approaches consider children's fiction (Freestone and O'Toole 2016), arts (Burke and Cutter-Mackenzie 2010) and places (Duhn 2012).

There is also growing recognition that 'environmental learning' can occur in formal and informal contexts and that 'it is possible to distinguish between environmental learning that is intended and planned and that which is not, such as learning that occurs naturally as a child explores their own environment' (Hacking, Scott, and Barratt 2007, 535). This calls again our attention to the significance of local communities as locus of belonging and participation. This is not a new topic in pedagogy, as John Dewey has long stressed how 'genuine education comes about through experience' even if 'experience and education cannot be directly equated to each other' ([1938]1997, 25). In a report on education in various countries, Malone (2008) explores outdoor learning in the local community (schools ground, natural parks, art galleries) to conclude that children involved in learning outside the classroom show gains in terms of cognitive, physical, emotional, social and personal skills, and thus suggesting that outdoor learning contributes to the development of the whole child (5–6). In another project focused in the context of landscapes in five European countries, Nilsson and Hensler (2001) recognize this holistic potential and argue for the promotion of three different and intertwined dimensions: 'heart – values, attitudes and personal qualities', 'hands – skills for life', and 'head – increasing knowledge' (10–11).

On a similar trend, Williams and Brown (2012) advocate for the creation of school gardens that might 'provide vibrant and meaningful learning sites to engage with the ideas and practices of sustainability' (x). In the US, Pierce (2015) uses actor-network theory as a guiding model for using school gardens with pupils, teachers and communities. He considers that school gardens can provide 'an alternative model for learning about plants and human relations with them' (463) by bridging schools and communities and promoting an eco-literacy that integrates a political dimension. This is, again, an essential element for a critical ecopedagogy since, in spite of the emphasis on the 'interconnection and reciprocity' (Gruenewald 2003, 34) with non-human spaces and beings, there might be a risk of using school gardens as yet another controlled, walled experience of 'nature'.

The work of Änggård (2010), based on an ethnography in a Swedish preschool, shows how, from the child's point of view, 'nature' can have different meanings:

as a classroom where children learn about nature in different ways; (…) as a home – a peaceful place in which to eat, sleep, socialize and play; (…) [and] as an enchanted world – a fairyland populated by fairy figures and animals with human traits. (10)

This study reminds us that we have to acknowledge children as active agents in learning (Corsaro 2005; Ferreira 2010; James and Prout 1990) – instead of mere recipients of the adults' pedagogical intentions and outcomes. Hacking, Scott, and Barratt (2007) advocate the right of children to be involved in decisions regarding environmental problems, and suggest that children research their local communities, share ideas about what they have found and contribute to problem-solving both in the present and the future. Isabel Menezes also defends that environmental education should avoid the imposition of meaning-making frameworks, but instead create opportunities for pupils to develop more integrated, complex and flexible visions of the self and the world that are decisive for their action-in-context (2004).

The recognition of children's agency also has significant epistemological consequences for environmental research and point to the need to actually involve children in discussing their views on nature, environmental problems and possible solutions (Caiman and Lundegård 2014; Mackey 2012). In this paper, our goal is to explore children's perspectives on *what*, *where* and *how* they learn about nature and what environmental problems they recognize in their urban communities, as well as possible solutions. Previous studies on children's conceptions of nature have mainly used interviews, drawings or observations to grasp children's understandings (e.g. Aguirre-Bielschowsky, Freeman, and Vass 2012; Caiman and Lundegård 2014; Mackey 2012; Myers, Saunders, and Garrett 2004), but have not involved children in group discussions regarding how they learn about nature and whether these experiences foster their environmental awareness and agency.

Methodology

Context

In Portugal, the Directorate General for Education (DGE) defines 'environmental education for sustainability' as a domain of citizenship education with particular relevance for contemporary societies. Citizenship education should be infused across the curriculum (e.g. subjects, activities and projects) from pre-school to secondary education (Decree-Law No. 139/2012, of 5 July, with the amendments introduced by Decree-Law No. 91/2013, of 10 July). Nevertheless, besides these general recommendations there are no specific mandatory guidelines for schools – even if environmental education in the school, and particularly in pre-school and primary school, has been essential for the growing awareness of environmental problems and the transformation of actual behaviours in the society as a whole (Guerra, Schmidt, and Nave 2008). In fact, pre-school and primary school teachers and children have been influential in encouraging the sorting of waste and recycling as family habits.

The schools involved in this research were intentionally selected as being representative of different pedagogical approaches regarding education, specifically environmental education. To ensure confidentiality all schools were renamed. With one exception, detailed below, most children in these schools are from Portuguese background.

The 'Green Kindergarten' (Green-K) is a public pre-school in a residential area on the outskirts of a large city, which was established as an eco-school in 2009. The eco-schools program is an international project of the Foundation for Environmental Education that supports schools in developing environmental education involving not only curricular but also co-curricular activities and schools interface with the community. Green-K is also involved in the environmental program promoted by SUMA, a public company in the field of waste collection, urban cleaning, and waste treatment management. The topics addressed by the school include water, residues and energies and involve not only in-class activities with the children and their parents, but also outdoor activities in the school's garden and vegetable garden. The school is actively involved in the collection of waste materials for recycling.

The 'Blue Sea College' (Blue Sea-C) is a private kindergarten and primary school, relatively close to Green-K. It is also an eco-school but only since 2015, following the guidelines of these schools. It is a bilingual (Portuguese and English) school. The school's educational project revolves around 'humanities, people and cultures', and has a strong environmental focus that includes approaching local and global environmental issues mainly in-class (e.g. global warming).

The 'Tree School' (Tree-S) is a private kindergarten situated in a nearby forest area close to the capital city. It presents itself as valuing outdoor education and project-based pedagogy, and assumes two models of environmental education: the forest school and Waldorf schools. The children at this school are from a multicultural background, with children from different European backgrounds (English, Spanish, German). The school's pedagogy values children's autonomy and freedom as well as experiential learning. The classrooms are small, but the outdoor space is very large with trees (pines, fruit trees), vegetable garden, and animals (rabbits, chickens …). There are waterproof clothes and galoshes available for all children, so that they can go outside and play even if it's raining.

The 'Woods School' (Woods-S) is situated in a village in a semi urban area and works under a model of homeschooling; parents have established an association that runs the project in the grounds of an old public primary school with the daily coordination of a mother who manages a diversified profile of weekly activities performed by volunteer teachers and school employees. The school's menu is vegetarian and the main pedagogical inspiration comes from Krishnamurti and Waldorf. The parents have high levels of environmental awareness and commitment, and the school's goal is to promote the integral development of the child through a variety of daily activities that involve meditation, theatre, and music. Daily the children go to walk on the nearby wood in silence – this activity is called the 'silent walk' [in English].

Data collection

In each school, we conducted a focus group with volunteer children, with the goal to empower them to speak out in their own words and encouraging them to voice opinions (Cohen, Manion, and Morrison 2007). Focus groups have the advantage of reproducing 'a group dynamic that is particularly familiar to the child who attends educational settings (…) [promoting] a safe environment through the presence of peers (…) [and] a balance of power between adults and children' (Dias and Menezes 2014, 254). Moreover, in topics with obvious social and political relevance, such as the environment, the use of focus groups seems particularly appropriate.

Ethical issues were central in this study, and permissions were obtained from the school boards and education professionals, as well as the parents and the children themselves. The groups were audio and videotaped, and the videos were later destroyed as negotiated with parents, after the transcription of the focus groups. The names of children mentioned in this paper are fictional. A total of 31 children participated in the groups, with an age range from 4 to 10 years old; there were 14 girls and 17 boys; most children were 5 years old, and the older group came from the Woods-S where the age range was from 6 to 10 years.

In order to foster children's expression, the first author presented herself as a researcher who wanted to know about children's views about nature. The introduction of the focus group used guided imagery to stimulate children's imagination, combined with relaxation and a background music of a gentle breeze to suggest a place outdoors. The instruction included the following:

> Now, we are all going to travel to a place in nature that we know or would like to know. … Let us close our eyes, if we want, and put our hands on our tummy and inflate it like a balloon, very gently. … [changing from we to you] Now, imagine yourself outdoors and start exploring this place … you see many things … what sounds do you hear? … What do you like most about this place? Is something bothering you? … What are you doing there? … Now, let's come back: open your eyes, stretch your arms and legs … Tell me, what did you imagine?

As the children evoked their imagined experiences, the researcher initially explored their views on nature and their awareness of environmental problems and possible solutions, and then proceeded with a discussion about their learning experiences about nature, in and out-of-school. In the end, the children were asked to draw their views of the main topics discussed in the group. The goal was to involve children in the definition of the main categories that emerged from the discussion; some children focused on the imagery exercise, others on their favorite theme from the discussion. They all explained individually to the researcher what their drawings meant and the researcher registered their views that were later used for the definition of the system of categories.

Data analysis

After the transcription of the focus groups, we performed a thematic analysis that explicitly combined inductive and deductive approaches, so that the categories were both 'data and theory-driven' (Braun and Clarke 2006, 18), and took into account how children themselves have interpreted the major topics of the discussion. In the first phase we did a repeated reading of our data, actively searching for meanings. Secondly, we coded the material. In the third phase we collected themes and sub-themes, which were revised, in order to 'form a coherent pattern' (20) in the fourth phase, and named and defined in a few interactions in the fifth phase. Finally, a report was produced (sixth phase), with excerpts that 'capture the essence of the point' (23).

Results

Images of nature

When describing what they saw in their imagery exercise, children referred to animals, plants and places that ranged from the wild to the familiar, and to mythical figures. Mostly, the co-actors of the imagined scenes were *animals* and *plants*. There were plenty of references to flowers and butterflies, but also to specific animals, including some that were absent, as in this discussion in the Blue Sea-C:

Antonio (5 years):　I was in a beach in the Algarve.

Interviewer:　Do you like to go there?

Antonio (5 years):　Yes, every day when I go to the beach (…). We go by boat and then I ask my mommy to eat a cake.

Interviewer:　And you like to sail in the boat?

Antonio (5 years):　I like to see the waves and to sail, but I do not see sharks!

Bruno (5 years):　The shark would eat you. There are no sharks there, Antonio!

Lucas (5 years):　There are sharks in the oceanarium.

In some cases, animals are seen almost as peers, as in Bruno's account of crabs keeping him company in the beach. As such, animals appear to be an essential element of nature, either because they are dangerous, but also because they are interesting and have clever behaviours, as shown in the interaction between the older children at the Woods-S:

Roberto (9 years):　I imagined I was in the Amazon …

Interviewer:　Have you ever been there?

Roberto (9 years):　No, but I would like it very much, and Mariana knows why!

Mariana (10 years):　I … I do not want to be eaten by a boa or a panther! Neither to be bitten by poisonous mosquitoes!

Roberto (9 years):　But it's not because of that! I am talking about flying serpents that fly from the trees and fall on top of animals and people. I would like to see them. These serpents have to make many mathematical calculation as they have to compute at what speed people or animals are passing by and they have to calculate when to throw themselves over them. If the person is running, it must be more difficult for them to calculate.

Andre (8 years):　If the person is running, the serpents have to calculate at what speed they are running. To throw over the person and get her.

Ricardo (7 years):　But the serpent can fall on the ground.

Lia (8 years):　And then what happens?

Andre (8 years):　It dies …

Roberto (9 years)　No … that does not happen.

(Laughs).

The places they 'went to' during the imagery include both *familiar and distant natural contexts*. There are mentions to the school grounds, to the zoo or the oceanarium, but also to gardens and forests in their community. As expected, children do not limit their visions of nature to their immediate surroundings (e.g. Aguirre-Bielschowsky, Freeman, and Vass 2012), and refer also to more distant places such as the beach, natural parks or the tropical forest and even, in the case of 6 year-old Dinis, 'for me, nature is things about space … stars, planets … and the Earth'. However, in some cases, anything green can count as nature as these children from the Green-K reveal:

David (5 years):   I was in a football pitch, with fresh grass, it was very cool and it was not dirty … everything was clean.

Francisco (5 years):   I was also in a pitch with clean and wet grass … because this way players could score more goals.

Luis (5 years):   I was also on a football pitch trying to catch a ball, I was playing football.

Another example comes from Rute's definition of nature that, as the following interaction shows, prevails in spite of other children's objections. It is important to note that Rute's kindergarten, Tree-S, has a vegetable garden and that she particularly likes going there to pick vegetables for the school's soup:

Rute (5 years):   Nature is soup.

Interviewer: Can you tell us why Rute?

Rute (5 years):   Because it has grass.

Maria (5 years):   It's not grass, it's spinach!

Ruben (4 years):   It's nettles!

Rute (5 years):   No, that stings. It's green beans.

[other children start giving examples of vegetables used for cooking soup]

Rute(5 years):   And we can also have one little carrot and grass. Because grass is good for your health and so are flowers.

(Laughs).

Dinis (6 years):   I prefer soup with carrots.

In many cases, nature is the outdoor place where you can play with devices such as slides, tunnels, swings or bicycles. Imagining yourself 'being in the nature' involves 'riding a bike by the sea' (Ricardo, 5 years) or picturing a playground where you have access to constructed play sets as with this group from the Green-K:

Rita (5 years):   I was in a park [in Portuguese, you can use the same word for playground and park], with slides and bars.

Interviewer: Did that park have nature?

Rita (5 years):   Yes, it had flowers, and grass and plants.

Miriam (6 years):   I imagined I was in a very cool house. It had flowers in the ceiling, it had flowerpots and it also had flowers in vases.

Joana (5 years):   I imagined I was on the park with grass, trees, and swings … I was riding a swing …

Alice (5 years):   I was also on a park with many flowers and grass, it had two swings and a tunnel.

However, while being in nature children recognize that *they can do many things*, as this group recalls their experience in a public park near the Tree-S:

Marta (5 years):   Yes, Monserrate has a castle and a lake with fishes. There is a fish that is very big and long … and very brown. We give him small pieces of bread to eat and he likes it.

Lucas (5 years):   Me and Dinis and Andre climbed a very big tree … it had many branches and giant roots … and we climbed all the way to the top. Then we lie on a branch that is like this … [he makes the gesture with his arm, meaning that the branch is horizontal].

Interviewer: And what else did you do?

Ana (6 years):   In Monserrate we do picnics … we sleep on the grass, we run.

Rute (5 years):   We run, we jump, we step on the grass, we climb up the trees …

Pia (4 years):   I like to play there with my friend.

Marta (5 years):   We play the princesses.

Nevertheless, being outdoors also implies being careful 'not to hurt yourself' (Maria, 6 years) and not to hurt the plants or the animals because 'if you step on a flower, it dies' (Bruno, 5 years).

The presence of *magic or mythical figures* is also central, suggesting continuity between the real world and fairyland. This magic has, however, a clear protective or caring goal. Raquel (5 years) saw 'little fairies (…) that made dresses, took care of the garden, planted seeds and watered the plans'. For Maria, from Blue Sea-C, the role of magic is protecting nature:

Maria (5 years):   I saw flowers, bees and a garden.

Interviewer: Do you enjoy being in that garden?

Maria (5 years):   Yes, but it was a magical garden.

Interviewer: Why was it magical?

Maria (5 years):   Because it had many butterflies and some special flowers that I could not pick.

Interviewer: And why not?

Maria (5 years):   So that people could not damage nature.

This protective disposition is no doubt related to the observation of several *environmental problems* that children witness directly in their walks in the community (littering is the most mentioned example), either with their family and the school. They also appear attentive to problems mentioned in the news, in children's TV programs. And not only are they aware that there are laws that forbid littering, and they have learned about separating waste and recycling in school, but also they can imagine creative solutions to problems, such as marine pollution as this group from Blue Sea-C:

Interviewer: And what do you think happens to the sea animals?

Several children: They get sick … really sick … they eat the garbage.

Isabel (5 years):   … and they die if they eat garbage.

Bruno (5 years):   And how can you do to save them?

Paulo (5 years):   They have to make a barracks underwater.

Interviewer: You would do a barracks under the water to clean the sea?

Several children: Yes … yes …

Paulo (5 years):   With equipment, some fins and a pair of glasses. Like divers.

On the whole, this diversity of topics was also present in the drawings children made to summarize the group discussions, with the playground and the garden emerging as the most frequent places of interest. If we look at other commonalities in the discussions, there was a tendency for the expression of a significant emotional connection with nature, even if ambivalent feelings (enjoyment/calm vs. fear/danger) also emerge. This generated mainly a protective disposition towards nature, either performed by children themselves, significant others or magical figures and features in nature itself. Caring for and taking care of nature appears clearly as a moral imperative, with strong criticism regarding human actions that result in menacing nature and animals.

Learning about nature

Many children refer to *learning experiences within families.* Children from the Green-K talk of significant daily experiences of taking care of nature with parents and grandparents and Francisco (5 years) describes how he and his mom reused materials for his Carnival prop:

Francisco (5 years):   To make the sword of my Carnival's costume, I transformed something old into something new. The costume did not have a sword, I made it.

Interviewer: Did you make the sword with reused materials?

Francisco (5 years):   Yes, with rubber bands, with paper, with a bag, my mother used a piece from the butter package and put it on plastic paper and glued everything. And now I have a sword.

As expected, *schools also emerge as a significant context for learning about nature,* and here a diversity of experiences emerge. In some cases, as in Woods-S, there are daily routines to promote communion with nature, with an emotional focus:

Mariana (10 years):   Every day, we go out for a walk, whether it is raining or sunny We go to a forest over there

Andre (8 years):   It's the silent walk.

Mariana (10 years):   And we talk as we go and come back in silence. But today, I, Lia and Ricardo went in silence and returned talking.

Lia (8 years):   We switched. (…)

Interviewer: But why do you do that in silence? Can you explain that to me?

Lia (8 years):   Not to waste energy.

Mariana (10 years):   To be in contact with nature.

Andre (8 years):   When I do the silent walk I stay focused.

Mariana (10 years):   And you can better observe around you everything that goes on in nature.

At Tree-S, children also demonstrate the emotional significance of experiential activities that involve a direct contact with nature, as Rute (5 years) mentions their daily visits 'to the trees outside, climbing up and down and picking the oranges'.

In fact, the schools' major contribution appears to be fostering the knowledge and practice of environmental activities: expressions like 'I know' and 'in school, we learn …' are frequent, suggesting a main cognitive focus of learning. This can involve study visits, but appears to be clearly infused in the curricula. For instance, at Green-K:

Cristiano (5 years):   We bring bottle caps and corks.

Joana (5 years):   And cereal boxes to make school projects.

Miguel (5 years):   I've learned in school that we should use things three times. Reuse things. (…)

Joana (5 years):   I throw the garbage into the recycling bin.

Cristiano (5 years):   And we have to put the garbage in the right recycling bin.

Interviewer: And tell me … do you know what they are?

All the children: I know all of them! Me too …

Interviewer: Then tell me …All the children [using the Portuguese popular names for the different bins] – Paper, glass, plastic and metal, batteries, electrical appliances …

Maria (6 years):   And the Wippy [to collect used clothing].

Generally, children mainly recognize learning about nature in the family context, in school and through the media, in this case TV, with references to advertising and children's programs. It is particularly interesting to notice, however, that while references to learning with families seem to involve mainly action and caring about nature, schools focus more on cognitive dimensions than on affective components of learning, with education *about* the environment clearly prevailing over education *in* the environment. The affective and experiential emphasis only emerges in the schools where alternative environmental educational models prevail. Nevertheless, in all cases there are no references to pedagogical experiences that involve children in approaching the civic or political dimensions of environmental learning.

Discussion and implications

Overall, our data echoes existing research in the field (e.g. Adams and Savahl 2015, 2016; Änggård 2010; Bonnett and Williams 1998; Stokas et al. 2016) as children express positive emotions towards nature, knowledge and awareness of environmental problems in their mostly urban communities and beyond, and meaningful experiences of interaction with and within natural environments that appear to foster a caring and protective attitude towards nature. Even if our data comes from a multiple-case study with a limited number of participants it is nevertheless interesting to note that the emotional and experiential components of learning, two strongly interrelated dimensions, seem to be less relevant in the schools with a more traditional approach to environmental education. These schools are representative of the 'cognitive knowing' focus' (Nazir and Pedretti 2015) of environmental education that prevails in Portuguese pre-schools and primary schools. The emphasis on emotional and experiential learning is more significant in the two schools, both private, that implement alternative eco-pedagogies. While all four schools seem to foster the children's sense of belonging and emotional commitment that are important components of sense of community, the promotion of inclusion with the natural environment seems to be recognized only by schools with a focus on eco-pedagogies. In all cases, however, acknowledging children's agency and capacity to influence decisions – another significant component of a sense of community – seems less evident in the children's accounts of their experiences. This finding suggests that more has to be done to recognize children's right to have a say regarding environmental problems and to foster their empowerment in promoting the sustainable development of their communities – an essential element, as Paulo Freire argued, of 'any educational practice of radical, critical or liberating character.' (2000, s/p).

   For most children nature in urban areas is associated with a playground, a finding that resonates with other studies (e.g. Stokas et al. 2016). Children report outdoor experiences mainly in these contexts, thus reinforcing the idea that current rearing practices are provoking a detachment and isolation from natural spaces that runs the risk of generating an artificial cognitive representation of the natural world (Wilson 1996), like when a child appears to value what is missing in his/her experiences – sharks, for instance – as much as what is actually there (the sea, the waves). Another consequence is the apparent significance of episodes vicariously experienced, either from the media or in the lives of significant others.

   Therefore, it appears that Krasny and Tidball (2009) call for a civic ecology approach that involves citizens, in cooperation with organized groups and governmental bodies, in restoring nature in cities through hands-on community projects, makes particular sense here. The participatory and interactional elements of these approaches have the potential to favor significant experiences with humans and non-humans that involve relations with a diversity of 'others' in terms not only of biodiversity (e.g. different species), but also from a social perspective (e.g. social class, ethnicity, expertise) that can foster personal, community and socio-ecological benefits (Hinchliffe and Whatmore 2006; Jagger, Sperling, and Inwood 2016; Krasny and Tidball 2009).

   However, a final note is necessary regarding an obvious ambivalence: while at the level of educational policy environmental education is frequently conceived as a dimension of citizenship education, a civic-political dimension seems to be absent from the discourses of the children. This can be a sign of the a-political tendency in the education of small children, viewed as innocent beings for whom politics is too complicated and conflictive. Therefore a call for a political environmental education is also a call for 'rethinking childhood' (Duhn 2012, 21). Obviously, children have been historically treated as ignored citizens (Bath and Karlsson 2016), while more and more educationalists advocate the need to view children as citizens here and now (Biesta and Lawy 2006; Dias 2012; Ferreira 2010; McCowan 2009; Ribeiro, Caetano, and Menezes 2016; Sarmento et al. 2009). As defended by Hacking, Barratt and Scott, a committed environmental citizenship implies 'overlaps between environmental education, citizenship education, and political education, [and that educational institutions and communities] provide opportunities for promoting children's participation as environmental stakeholders' (2007, 534). The challenge is therefore to overcome the tendency of contemporary societies, as Hannah Arendt would say, to exclude children 'from our world and leave them to their own devices' (Arendt [1954] 1961, 196), denying their potentially central role in 'the task of renewing a common world' (196).

On the other hand, our research also suggests that there is no sign of granting an expansion of citizenship and politics to nonhumans by advocating for 'a more than human condition' (Asdal, Druglitrø, and Hinchliffe 2016). We seem to be facing two potentially oppressive dichotomies – children vs. adults, humans vs. nonhumans – that undermine the transformative potential of these pedagogies. This surely implies exploring ways to meaningfully include children's visions, concerns and ideas as well as to directly engage them with the nonhumans around them (Jagger, Sperling, and Inwood 2016; Krasny and Tidball 2009; see also Rautio, this issue) – and if there is a clear potential in the civic ecology approaches mentioned above, their relational and participatory elements are not inevitable remedies for solving these dichotomies. Promoting a political ecoliteracy (Gruenewald 2003; Pierce 2015) depends on having children critically examine how decisions about their communities implicate on the quality and sustainability not only of the natural environment but also of democracy itself, and this implies creating school and community spaces where this political (inter)action – between children and adults, between humans and non-humans – is put into practice.

Acknowledgements

The authors wish to thank José Pedro Amorim, Teresa Silva Dias and Tiago Neves for their useful comments and support regarding this research.

Disclosure statement

No potential conflict of interest was reported by the authors.

ORCID

Isabel Menezes (iD) http://orcid.org/0000-0001-9063-3773

References

Adams, S., and S. Savahl. 2015. "Children's Perceptions of the Natural Environment: A South African Perspective." *Children's Geographies* 13 (2): 196–211.
Adams, S., and S. Savahl. 2016. "Children's Discourses of Natural Spaces: Considerations for Children's Subjective Well-Being." *Child Indicators Research*. doi:10.1007/s12187-016-9374-2.

Aguirre-Bielschowsky, I., C. Freeman, and E. Vass. 2012. "Influences on Children's Environmental Cognition: A Comparative Analysis of New Zealand and Mexico." *Environmental Education Research* 18 (1): 91–115.

Albanesi, C., E. Cicognani, and B. Zani. 2007. "Sense of Community, Civic Engagement and Social Well-Being in Italian Adolescents." *Journal of Community & Applied Social Psychology* 17: 387–406.

Änggård, E. 2010. "Making Use of 'Nature' in an Outdoor Preschool: Classroom, Home and Fairyland." *Children, Youth and Environments* 20 (1): 4–25.

Arendt, H. [1954] 1961. "The Crisis in Education." In *Between past and Future: Six Exercises in Political Thought*, edited by H. Arendt, pp. 173–196. New York: The Viking Press.

Asdal K., Druglitrø T., and Hinchliffe S. J. 2016. "Introduction: The 'More-than-Human' Condition. Sentient Creatures and Version of Biopolitics." In *Humans, Animals and Biopolitics: The More-than-Human Condition*, edited by K. Asdal, T. Druglitrø and S. J. Hinchliffe, pp. 1–29. Abingdon, Routledge.

Bath, C., and R. Karlsson. 2016. "The Ignored Citizen: Young Children's Subjectivities in Swedish and English Early Childhood Education Settings." *Childhood*, 1–12. doi:10.1177/0907568216631025.

Biesta, G. J. J., and R. Lawy. 2006. "From Teaching Citizenship to Learning Democracy: Overcoming Individualism in Research, Policy and Practice." *Cambridge Journal of Education* 36 (1): 63–79.

Bonnett, M., and J. Williams. 1998. "Environmental Education and Primary Children's Attitudes towards Nature and the Environment." *Cambridge Journal of Education* 28 (2): 159–174.

Braun, V., and V. Clarke. 2006. "Using Thematic Analysis in Psychology." *Qualitative Research in Psychology* 3 (2): 77–101.

Burke, G., and A. Cutter-Mackenzie. 2010. "What's There, What If, What Then, and What Can We Do? An Immersive and Embodied Experience of Environment and Place through Children's Literature." *Environmental Education Research* 16 (3-4): 311–330.

Caiman, C., and I. Lundegård. 2014. "Pre-School Children's Agency in Learning for Sustainable Development." *Environmental Education Research* 20 (4): 437–459.

Cohen, L., L. Manion, and K. Morrison. 2007. *Research Methods in Education*. Oxon: Routledge.

Corsaro, W. A. 2005. *The Sociology of Childhood*. Thousand Oaks, CA: Sage.

Dewey, J. [1938] 1997. *Experience and Education*. New York: Touchstone.

Dias, T. S. 2012. "Como Pensam 'Elas' a Organização Das Sociedades E O Exercício Da Cidadania? Do Desenvolvimento Do Pensamento Político À Vivência Da Cidadania Participada Em Contexto Escolar No Pré-Escolar E Ensino Básico." PhD diss., University of Porto.

Dias, T. S., and I. Menezes. 2014. "Children and Adolescents as Political Actors: Collective Visions of Politics and Citizenship." *Journal of Moral Education* 43 (3): 250–268.

Dockett, S., E. Kearney, and B. Perry. 2012. "Recognising Young Children's Understandings and Experiences of Community." *International Journal of Early Childhood* 44: 287–305.

Duhn, I. 2012. "Making 'Place' for Ecological Sustainability in Early Childhood Education." *Environmental Education Research* 18 (1): 19–29.

Ferreira, M. 2010. "'- Ela É a Nossa Prisioneira!': Questões Teóricas, Epistemológicas E Ético-Metodológicas a Propósito Dos Processos De Obtenção Da Permissão Das Crianças Pequenas Numa Pesquisa Etnográfica." *Revista Reflexão E Acção* 18 (2): 151–182.

Ferreira, P. D., J. L. Coimbra, and I. Menezes. 2012. "'Diversity within Diversity' – Exploring Connections between Community, Participation and Citizenship." *Journal of Social Science Education* 11 (3): 120–134.

Freestone, M., and J. M. O'Toole. 2016. "The Impact of Childhood Reading on the Development of Environmental Values." *Environmental Education Research* 22 (4): 504–517.

Freire, P. 2000. *Pedagogia Da Indignação: Cartas Pedagógicas E Outros Escritos*. São Paulo: Editora UNESP.

Gruenewald, D. A. 2003. "At Home with the Other: Reclaiming the Ecological Roots of Development and Literacy." *The Journal of Environmental Education* 35 (1): 33–43.

Guerra, J., L. Schmidt, and J. G. Nave. 2008. "*Educação Ambiental Em Portugal: Fomentando Uma Cidadania Responsável.*" Accessed March 29 2016. http://www.aps.pt/vicongresso/pdfs/681.pdf

Hacking, E. B., W. Scott, and R. Barratt. 2007. "Children's Research into Their Local Environment: Stevenson's Gap, and Possibilities for the Curriculum." *Environmental Education Research* 13 (2): 225–244.

Hedefalk, M., J. Almqvist, and L. Östman. 2015. "Education for Sustainable Development in Early Childhood Education: A Review of the Research Literature." *Environmental Education Research* 21 (7): 975–990.

Hinchliffe, S., and S. Whatmore. 2006. "Living Cities: Towards a Politics of Conviviality." *Science as Culture* 15 (2): 123–138.

Jagger, S., E. Sperling, and H. Inwood. 2016. "What's Growing on Here? Garden-Based Pedagogy in a Concrete Jungle." *Environmental Education Research* 22 (2): 271–287.

James, A., and A. Prout. 1990. *Constructing and Reconstructing Childhood: New Directions in the Sociological Study of Childhood*. Oxford: Routledge.

Krasny, M. E., and K. G. Tidball. 2009. "Community Gardens as Contexts for Science, Stewardship, and Civic Action Learning." *Cities and the Environment* 2 (1): article 8, 18. http://escholarship.bc.edu/cate/vol2/iss1/8

Latour, B. 2004. *Politics of Nature: How to Bring the Sciences into Democracy*. Cambridge, MA: Harvard University Press.

Machemer, P. L., S. P. Bruch, and R. Kuipers. 2008. "Comparing Rural and Urban Children's Perceptions of an Ideal Community." *Journal of Planning Education and Research* 28: 143–160.

Mackey, G. 2012. "To Know, to Decide, to Act: The Young Child's Right to Participate in Action for the Environment." *Environmental Education Research* 18 (4): 473–484.

Malone, K. 2007. "The Bubble-Wrap Generation: Children Growing up in Walled Gardens." *Environmental Education Research* 13 (4): 513–527.

Malone, K. 2008. "Every Experience Matters: An Evidence Based Research Report on the Role of Learning outside the Classroom for Children's Whole Development from Birth to Eighteen Years." Report commissioned by Farming and Countryside Education for UK Department Children, School and Families, Wollongong, Australia. Accessed December 28 2015. http://www.face-online.org.uk/docman/news/every-experience-matters/download

McCowan, T. 2009. *Rethinking Citizenship Education: A Curriculum for Participatory Democracy*. London: Continuum.

Menezes, I. 2004. "Ambiente E Transversalização Curricular: Potencialidades E Limites Da Educação Ambiental Na Escola." *Educação, Sociedade & Culturas* 21: 130–150.

Myers Jr., O. E., C. D. Saunders, and E. Garrett. 2004. "What Do Children Think Animals Need? Developmental Trends." *Environmental Education Research* 10 (4): 545–562.

Nazir, J., and E. Pedretti. 2015. "Educators' Perceptions of Bringing Students to Environmental Consciousness through Engaging Outdoor Experiences." *Environmental Education Research* 22 (2): 288–304.

Nilsson, A.-C., and M. Hensler, eds. 2001. *Outdoor Education: Authentic Learning in the Context of Landscapes. a Transnational Co-Operation Project Supported by the European Union*. Kisa: Kinda Education Center.

Pierce, C. 2015. "Against Neoliberal Pedagogies of Plants and People: Mapping Actor Networks of Biocapital in Learning Gardens." *Environmental Education Research* 21 (3): 460–477.

Pooley, J. A., L. T. Pike, N. M. Drew, and L. Breen. 2002. "Inferring Australian Children's Sense of Community: A Critical Exploration." *Community, Work & Family* 5 (1): 5–22.

Ribeiro, A. B., A. Caetano, and I. Menezes. 2016. "Citizenship Education, Educational Policies and NGOs." *British Educational Research Journal* 42 (4): 543–728.

Sarmento, T., F. I. Ferreira, P. Silva, and R. Madeira. 2009. *Infância, Família E Comunidade: As Crianças Como Actores Sociais*. Porto: Porto Editora.

Stokas, D., E. Strezou, G. Malandrakis, and P. Papadopoulou. 2016. "Greek Primary School Children's Representations of the Urban Environment as Seen through Their Drawings." *Environmental Education Research*. doi:10.1080/13504622. 2016.1219316.

Williams, D., and J. D. Brown. 2012. *Learning Gardens and Sustainability Education: Bringing Schools and Community to Life*. New York: Routledge.

Wilson, R. A. 1996. "The Development of the Ecological Self." *Early Childhood Education Journal* 24 (2): 121–123.

'Staying with the trouble' in child-insect-educator common worlds

Fikile Nxumalo and Veronica Pacini-Ketchabaw

ABSTRACT

Classroom pet programs have become extremely popular in urban North American early childhood classrooms. This article challenges anthropocentric child-pet pedagogies by proposing common world pedagogies of 'staying with the trouble.' Drawing from a common world multispecies ethnography in one early childhood centre, the authors engage with the specificities of educators' and children's everyday practices of caring for and detaching from an introduced species of Vietnamese walking stick insects. The paper argues that the child-pet-educator relations that emerged through these practices are a site at which to trace and disentangle commodified relations of enclosure and invasion in urban nature pedagogies within anthropogenically damaged places. We conclude by suggesting that classroom pet pedagogies need to enact a more-than-human relational ethics which subverts child development discourses and unsettles children and animals as innocent couplings.

Interrogating narratives of pets in early childhood

Walking stick insects have been a presence in this early childhood programs for years, since they were purchased by one of the programs from a local pet store. The insects' rapid proliferation became unmanageable for the programs' educators. Some chose to destroy their classroom stick insects; others gave them to children to take home as pets; still others gave them away to other early childhood programs. Yet the problem of what to do with these classroom pets continues.

Pets in the classroom are an explicit curricular component in many urban North American early childhood programs. Pets are introduced to teach children biological concepts and facts, such as animal cycles, habitats, feeding, and reproductive behaviours (Melson 2001). Classroom pets are also popular for developing compassionate and empathetic behaviours, morals, self-esteem, and responsibility in children, as well as for reducing stress (Jegatheesan and Meadan 2006; Meadan and Jegatheesan 2010; Myers 2007; Ruid and Beck 2000). These pedagogies fit with the dominant anthropormorphic and innocent framing of child-animal relations in North American literature and popular media (Taylor 2013; Timmerman and Ostertag 2011). Research on keeping classroom pets tends to be situated within child development, with most studies emphasising pets' benefits for enhancing children's scientific knowledge, social and emotional development, and socialisation with peers.

In recent years, providing care for classroom pets has been situated in relation to children's development of a caring connection with the nonhuman world (Louv 2008; Pelo 2009). In urban settings

Figure 1. Classroom Terrarium

encounters within the interrelated contexts of settler colonialism and anthropogenic environmental effects (Nxumalo 2015, 2016; Pacini-Ketchabaw and Nxumalo 2015).

The walking stick insects in this early childhood centre were originally purchased from a local pet store by educators in the after-school care program. Walking stick insects are marketed as appropriate and useful animals to teach children about caring for pets, and insect anatomy and behaviour (Locke 2009). Walking stick insects emerged as an issue for inquiry in our research when we noticed the contradictory and ambivalent affects these insects generated in the classrooms. Over a period of nine months, we had multiple conversations with educators and children, wrote weekly pedagogical narrations (Pacini-Ketchabaw et al. 2014) with the educators, and documented our thoughts and dilemmas in field notes. We became interested in the unevenly entangled child-insect-educator common worlds (Taylor 2013) as a means to interrupt the normative child-centred approaches to making meaning of child-pet relations in the classroom that initially brought the stick insects into these classrooms.

Enclosures

In these particular classrooms, walking stick insects have been kept in a terrarium on a shelf for three years now. The terrarium is covered with a wire mesh screen (Figure 1). Occasionally, a limb escapes the bounds of the screen. Some children are keen to touch the stick insects when this happens, while others draw back from the possibility of contact. Blackberry leaves from a nearby forest are regularly placed in the terrarium as food for the insects. A white paper towel at the bottom of the terrarium catches both waste and eggs.

Walking stick insects, including both the variety introduced to these classrooms, *Ramulus artemis*, whose original home is in the tropical forests of Vietnam, and the Indian walking stick *Carausius morosus*, which has also been introduced as a pet in North America, have emerged as one of the most popular types of insects to be kept as home and classroom pets (Locke 2009). *Carausius morosus* is also known as 'laboratory stick insect' due to its common use in scientific experiments ranging from finding

Figure 2. Child-Stick Insect Encounters

compounds with potential to address antibiotic resistance, to robotics engineering inspired by the stick insects' walking mechanisms (Ericson 2013; Locke 2009). Stick insects have even been taken into space for the purpose of conducting scientific tests on the effects of cosmic radiation and low gravity on their eggs (Locke 2009)! Stick insects have been studied for their size (they are among the longest insects), their ability to shed their skin and to remain camouflaged by mimicking the colour and other physical characteristics of tree branches and leaves, their ability to regenerate limbs, and their ability to reproduce by parthenogenesis, where eggs develop without male fertilisation (Alderton 1992; Amateur Entomologists' Society 2016; Keeping Insects 2016).

At times the walking stick insects' cling to the mesh cover, making them easy to see; at the same time their close proximity to each other and their ability to camouflage themselves amidst the blackberry branches, often makes it difficult to discern individual bodies. The children use magnifying glasses to see and know stick insects (Figure 2), and they often use a spray bottle to 'water' the insects to maintain humidity in their habitat through the mesh cover.

Our experimentations with common world pedagogies sparked a conversation about the difficult terrain we found ourselves in when we situated these classroom practices within commodified animal relations requiring capture, disposability, and enclosure (Gillespie and Collard 2015). Worlds of bodily and spatial enclosures, Collard (2014b) writes, are 'of deadening and growing sameness' (para. 9) because their intent is to regulate the (perceived) wilderness of animals. The colonial dualism of a pure 'wild' nature separate from human culture is what creates these spaces of enclosure and practices of control, 'conservation,' invasion-management, and captivity (Bell and Russell 2000; Collard 2012; Dempsey 2016; van Dooren 2015; Fawcett 2013). Moreover, animals such as stick insects are connected to relations of 'lively commodity' (Collard and Dempsey 2013) in which a multibillion-dollar industry in legal wildlife trade (and many more billions in illegal trade) exists, much of which consists of importing captured live animals as exotic pets. The pet trade, according to National Geographic (2016), 'along with the popular practice of framing their carcasses' (para. 6) threatens stick insects as wild species.

Collard and Dempsey (2013) note that for animals that are kept alive for the entire time they are commodities, being alive is part of their value while being dead has no value. Yet Collard (2014a) also warns that negotiating the fraught ethics of animal enclosure in the Anthropocene does not mean moving toward grand narratives of animal rights. This negotiation is the task we struggle with in this child-insect-educator ensemble.

We are inspired by the work of Rose and van Dooren (2017) to attend to this ensemble as a practice of becoming witness that requires

> an openness to others in the material reality of their own lives [as] noisy, fleshy, exuberant creatures with their multitude of interdependencies and precarities, their great range of calls, their care and their abundance along with their suffering and grief. (124)

Our task in the Anthropocene, Rose and van Dooren remind us, is to become attentive witnesses to these troubling enclosures and to respond to them through ongoing encounter, recognition, and curiosity that exceeds all rational calculation.

The children notice that several of the stick insects in the terrarium are missing their legs. They worry about this and wonder what we might be able to do. Not knowing why it happens and so being unable to answer the children's questions, we consult with biologists in our local community and consult environmental websites and books to become more attentive. We learn that stick insects in captivity need big cages with plenty of room because when the insects are overcrowded, they will bite or knock each other's legs off.

Focusing on the ecologies of these particular insects in this specific locality, we seriously considered what possibilities were foreclosed for these undomesticated pets living captive lives, and we traced the specific ways that colonial histories, practices of life-as-property, and histories of animal enclosure have unfolded in this early childhood centre. We wondered how stick insects from Vietnam had come to be in pet stores in British Columbia and then become such a popular feature of classrooms. We considered our own and the children's implicatedness in commodified pet relations that now seem such a normal part of urban classroom pedagogies. We discovered that little is known about how these particular insects came to be in North America. However, researchers link their arrival with European colonization given that 2273 insect and arachnid species have been introduced in North America since the first European settlers arrived (Iowa State University 2016).

An educator discusses with the children the enclosure of walking stick insects in the classroom terrarium. The children provide ideas: Can we release them into the wild? Can we bring them back to where they came from? Should we bring them to the pet store? We struggle to make sense of the notion of wilderness. Can we refer to these walking stick insects as 'wild' when they were hatched in the terrarium? Can there be a 'going back'? How do colonising, racialising, and anthropocentric constructions shape our questions and our practices?

Geographer Jamie Lorimer (2015) writes that the colonial imaginary of wilderness creates 'a pure and ahistorical place for Nature – natural by virtue of being untouched by human hands' (22). Thus nature is a space where humans are absent. Through a wilderness framework, however, animals are positioned as domesticated or undomesticated, and humans are recentred. A more productive concept than wilderness, Collard (2014b) suggests, is wildness. Within the politics of human-animal relations, wildness is both a critical concept and a radical alternative to the totalising narratives of the Anthropocene and of animal rights, and to the endlessly extended reach of capitalist exploitation through scientific testing, the exotic pet trade, and the global food industry. Collard, Dempsey, and Sundberg (2015) write:

> The degree to which an animal is wild … has little to do with its proximities to humans and everything to do with the conditions of living, such as spatial (can the animal come and go), subjective (can the animal express itself), energetic (can the animal work for itself), and social (can the animal form social networks). These are conditions of possibility, of potential, not forced states of being. (328)

Similarly, Lorimer (2015) refers to wildlife as a vernacular political concept that counters the idea of wilderness. Wildlife for him suggests processes, describes 'ecologies of becoming' (7), provokes 'curiosity, disconcertion, and care,' and demands 'political processes for deliberating discord among multiple affected publics' (11). In other words, by focusing on wildlife we can engage with the child-insect-educator

Figure 3. Stick Insect Drawing

ensemble as a political-ethical-ecological issue that requires us to confront the challenge of 'living with human and nonhuman difference' (11).

Twisted within enclosures, commodities, wilderness, wildness, wildlife, exotic pet trade, commodification, and so on in the context of classroom pet relations, we examine every issue that confronts us, even if our engagements offer no clear resolutions. How to move away from universalisms? How to create conditions for wildness, for wildlife, for both humans and more-than-humans in these classrooms? How to engage in critical engagements that are always contingent and situated materially, discursively, and affectively within the particularities of the common worlds that children, walking stick insects, and educators co-inhabit in this early childhood centre?

Invasion

Our initial question of how to manage and control the growing population of stick insects turns toward how to co-inhabit the classroom with stick insects. Yet we quickly realise that these two challenges are not necessarily exclusive. We are to respond to the rapid reproduction of these walking stick insects. Their rapid reproduction creates unlivable conditions for them in the classroom.

In 'the wild,' stick insect populations are controlled because even though the females often lay hundreds of eggs at a time, many predators (primarily bats) eat both the insects and their seed-like eggs. In captivity, however, with few predators (in this case us), stick insects proliferate rapidly (Alderton 1992). This rapid multiplication has led educators to discuss the threat of invasion these walking stick insects might present to other species we co-inhabit with in British Columbia's forests, particularly blackberry bushes, if the insects were to escape. We didn't find any information that suggests these insects are invasive in British Columbia's forests, but biologist Locke (2009) at the University of Alberta warns that in southern California, where the climate is mild and much closer to that of stick insects' native habitats, feral populations of walking stick insects have been causing problems for the last ten years.

Thinking with invasion is a way to remind us of the ways European settler colonialism has damaged and continues to damage forest ecosystems in the region now known as British Columbia. The challenge, though, as Frawley and McCalman (2014) and Lesley Head and her colleagues (2015) explain, is that is difficult to view approaches to controlling invasive species within a lens of eradication that not only signals a return to 'original landscapes' but also reproduces a colonial grand narrative of anthropogenic mastery. Frawley and McCalman (2014) suggest that we instead 'interrogate the complex and ongoing community concerns about invasive species and their ecological and cultural impacts that we will together have to face in a climate-changing world' (5). With this idea in mind, we ask: What are the possibilities for troubling meanings of invasion as an already known entity that is easily attached to practices of control and eradication (Head et al. 2015)? When juxtaposed with the problematics of commodified enclosure, what are the potentialities and the risks of troubling categories of invasion and noninvasion and looking instead at the naturecultures that children, stick insects, educators, and blackberry leaves, among others, are collectively and unevenly immersed in? Which approach to ethics might be most productive for co-inhabiting with these walking stick insects in this classroom?

Despite the educators' practice of removing the paper towel onto which the stick insect eggs drop, the walking stick population continues to grow. The process of destroying the eggs causes anxiety for the educators. The plethora of websites that provide information on keeping stick insects as pets often include instructions for controlling reproduction by freezing the eggs or destroying the walking stick insects by boiling them. In fact, freezing the eggs is recommended as an experiment for children to learn about the insects. John Locke's website provides step-by-step instructions for freezing eggs for periods of one hour to two weeks to test the effects of cold treatment on egg hatching. Locke (2009) notes: 'Cold treatment is only one of several 'environment treatments' that are possible. Think of some more and test them using this experimental system. In the 'Interesting Facts' section, see the effect of temperature on the appearance of male insects' ('Longterm #2' para. 4).

Some educators decide to freeze the eggs in a bag prior to discarding them in the garbage. Another reads that the eggs can be boiled, and this becomes another attempt to control the population boom in the classroom. These practices, which took place outside of the children's presence and view, did not always work. Several times when educators went to throw the frozen eggs out, they found that some of them had hatched and the insects were still moving.

Despite the periodic (and traumatic) removal and freezing of eggs and boiling of stick insects, the population keeps increasing, and the children and the educators continue to worry about the crowded conditions in the terrarium. Concretely, we are left with the decision of whether and how to kill (or let die) these walking stick insects we co-inhabit with.

In looking back at these processes of killing, we might consider how carrying out the acts of killing outside of children's presence foreclosed the possibility of a rich exploration of pain, suffering, loss, death and grief (Lloro-Bidart 2015; Russell 2017; Spannring 2017). We might perhaps also view these acts of killing as intense sites of ethical response as educators continually stayed with the difficult questions surrounding the pain and suffering of the stick insects. While the pain of another cannot be the sole focus of ethics, it is nonetheless capable of issuing a claim over us and therefore remains an important site of and for ethics. This might include arguing against generalized, abstracted sets of ethics, and instead supporting a situated ethics that is always subject to contestation and change (van Dooren 2015; Haraway 2008).

For instance, we might need to rethink some examples of the violent killing that is done in the name of conservation, and challenge the valuing of some animal lives over others based on, for instance, 'cuddliness and familiarity' or 'sentience' (van Dooren 2015, 24). Some form of preferential decision-making regarding killing is often unavoidable within spaces of conflicting ethics. However, it is also important to resist simplistic decision-making that does not consider carefully how certain animals come to be seen as deserving of care while others become easily and violently killable (in the name of conservation, for example) (van Dooren 2015). On the contrary, we take up the challenge of remaining attentive to the ethical obligations of the life and death contexts that we inhabit.

One educator is concerned that the children have lost interest in the stick insects, which have become an invisible backdrop in the classroom. In an effort to reengage children with the insects, she moves the tank to the centre of the room alongside paper and pencils as invitations to draw. The terrarium-stick-insect-paper-pencils beckon to the children. As they draw and engage in dialogue with each other, some notice that some of stick insects are still missing their legs. Several drawings (as illustrated in Figure 3) highlight the crowded multiplicity of stick-insect bodies in the terrarium. While concern-ambivalence-indifference (and more) might already be seen as inhabiting everyday encounters with the stick insects, the children's drawings point to a shift on the part of educators toward concern. While stick insects remain an object of learning for children (perhaps enacted in this drawing encounter as learning empathy toward more-than-human beings), a perceptible shift toward the actual bodies and precarious liveability of the stick insects emerges.

Caring-detachment

We were drawn to van Dooren's (2014a) discussion of Puig de la Bellacasa's (2012) concept of care, which sees caring as a situated, complex, undetermined, ambiguous, noninnocent, fraught, affective, contested, and compromised practice. van Dooren argues that caring is an affective state because it 'is an embodied phenomenon, the product of intellectual and emotional competencies'; an ethical obligation because it 'is to become subject to another, to recognise an obligation to look after another; and a practical labour because it 'requires more from us than abstract well wishing, it requires that we get involved in some concrete way' (291).

Yet, multispecies caring might also involve practices of detachment, including killing (Ginn 2014). Drawing on Haraway's work on companion species, Ginn (2014) reminds us that at times 'life's flourishing requires practices of exclusion and violence that are organised through hoped-for absence' (541). The point is not that 'some creatures are 'killable' by virtue of the circumstances into which they are birthed' (538), but that sometimes we are required to 'kill well' by 'meeting the ethical injunctions to be curious and to hold in mutual regard' (538). This is what Ginn calls ethical detachment – where detachment does not imply indifference, complete separation, or hostile withdrawal. He refers instead to an ethics that might emerge alongside, in negotiation with, as well as subsequent to practices of relating. In other words, the violence of some encounters with more-than-human others might be also termed 'caring' when considering that ethical modes of caring might also involve modes of detachment (Ginn 2014). Importantly, in this understanding, detachment does not imply a modernist reinscription of the human subject as outside of the world and able to enact an objective stance and complete separation (Haraway 1988). Instead, detachment here denotes affective, imperfect, consequential and caring acts of responding. In this classroom these practices, which involved killing stick insects, emerged out of grapplings with the ethics of educators' implicatedness and entanglement in the kinds of lives these particular insects were living. Ethical detachment, understood as complex and imperfect forms of caring, describes attempts to find better ways to kill the walking stick insects. Ginn (2014) writes that

> de-composition can be done with care: killing mindfully; avowing the violences of our actions; acknowledging the limits of our capacities to bend space to our will and imagination; being willing to recognise and be open to the vulnerability of non-human others and, perhaps, to be transformed by that recognition. (541)

As educators and children in our project learned to be affected by, to feel responsibility toward, and to struggle with which practices of detachment might be more ethical than others, they learned how to care within their common worlds, and 'meet the challenge of respecting and living with unwanted others' (Ginn 2014, 541).

From these understandings of detachment, the educators embodied practices of removing and killing eggs as a form of caring-detachment. Their acts of culling required them to juggle regret, resentment, guilt. They questioned who these acts of caring benefited – the children? the stick insects? themselves? They questioned the worth of keeping the stick insects amid the ongoing necessity of destroying their eggs. They questioned the environmental costs of releasing the stick insects to the forest, as some of the children had suggested. They questioned the differential ethics that made some species more easily killable than others. They questioned what living and dying well meant for the stick insects, and they struggled with the ethics of being the ones who must decide amid the impossibility of a morally pure position. They grappled with how 'best' to kill the insects to relieve their crowded living conditions: Should they boil the stick insects or freeze their eggs to death or perhaps withdraw food and water? What practices of 'mindful killing' and 'ethical detachment' might they enact in relation to the stick insects? One educator asked, 'Do we keep killing them little by little forever?'

By the end of the school year, all of the stick insects had been donated to an entomology research laboratory that had a large terrarium. By this action, stick insects became enrolled in unknown and perhaps problematic worldings of care within the research lab. And, while these walking stick insects are 'gone,' other stick insects are still present in other classrooms. As Puig de la Bellacasa (2012) reminds us, practices of care provide no smooth resolution.

Staying with the trouble in common world pedagogies

Framing the stick insect-child stories within common worlds framings allows us to challenge the nature/culture divides embedded within normative narratives of urban environmental education and to refocus our analysis to nature-culture entanglements (Taylor, this issue). Paying attention to the messy entanglements of urban-nature pedagogies also brings an attunement to the lives of nonhuman others in early childhood classrooms. It focuses on the complexities of educator-animal-child relations within lifeworlds, rather than solely on how children might benefit from these relations. In so doing, it interrupts anthropocentric understandings that view urban classroom pet relations through the lens of children's *learning-about* animals. Rather than focusing solely on a predetermined scientific lesson about stick insects, common world pedagogies allowed us to also engage in a contextual, situated walking stick pedagogy that paid attention to the place-specific temporalities of classrooms and to what it means, specifically, to live and learn with these particular 'introduced animals' in a space where ethics and the effects and affects of killing and making killable are all part of these modes of learning.

Thinking through concepts of invasion, wildness, wildlife, caring, and detachment, we brought an attunement to more-than-human relations that matter for the urgent environmental challenges children stand to inherit. These attunements challenge the framing of pet-child relations solely within the humancentric lens of children's development, which is insufficient at this time of anthropogenic change that calls for radically different ways of viewing our relationships with more-than-human others.

Staying close to the specific common worlds of children, stick insects, and educators in these particular classrooms and to their complicated belongings in British Columbian forests (and staying away from universalised to transcendent prescriptions), we grappled with the question of what it might mean (or look like) to respond recuperatively with children amid the 'assemblages of liveability and unlivability' (Tsing 2015, n.p.) that emerge from urban classroom pet pedagogies.

Our common world pedagogies are certainly not about helping children to arrive at clear answers. Rather, they are about staying with the trouble of recuperation by questioning capitalist practices of commodified enclosure that have brought the stick insects to this urban early childhood centre, and simultaneously paying attention to the affective, and imperfect, practices of care-detachment that are always present in settler colonial spaces in troubled times of anthropogenic damage.

Note

1. We use 'more-than-human' to refer to all life that is not considered human, with an understanding that humans are inextricably entangled and co-constituted with their more-than-human relations and responsibilities (Haraway 2008; Le Grange 2011; Whatmore 2006).

Disclosure statement

No potential conflict of interest was reported by the authors.

References

Alderton, D. 1992. *A Step-by-step Book About Stick Insects*. Neptune City, NJ: TFH Publications.
Amateur Entomologists' Society. 2016. "Stick Insect Care Sheet." http://www.amentsoc.org/insects/caresheets/stick-insects.html.
Bell, A. C., and C. L. Russell. 2000. "Beyond Human, Beyond Words: Anthropocentrism, Critical Pedagogy, and the Poststructuralist Turn." *Canadian Journal of Education* 25 (3): 188–203.
Bone, J. 2013. "The Animal as Fourth Educator: A Literature Review of Animals and Young Children in Pedagogical Relationships." *Australasian Journal of Early Childhood* 38 (2): 57–64.
Collard, R.-C. 2012. "Cougar–Human Entanglements and the Biopolitical Un/Making of Safe Space." *Environment and Planning D: Society and Space* 30 (1): 23–42. doi:10.1068/d19110.
Collard, R.-C. 2014a. "Wild. The Multispecies Café." [Companion Website to the Book of the Same Title, edited by E. Kirksey and published by Duke University Press]. http://www.multispecies-salon.org/wild/.
Collard, R.-C. 2014b. "Putting Animals Back Together, Taking Commodities Apart." *Annals of the Association of American Geographers* 104 (1): 151–165. doi:10.1080/00045608.2013.847750.
Collard, R.-C., and J. Dempsey. 2013. "Life for Sale? The Politics of Lively Commodities." *Environment and Planning A* 45 (11): 2682–2699. doi:10.1068/a45692.
Collard, R.-C., J. Dempsey, and J. Sundberg. 2015. "A Manifesto for Abundant Futures." *Annals of the Association of American Geographers* 105 (2): 322–330.
Common World Childhoods Research Collective. 2015. "Children's Relations with Place, with Materials, and with Other Species." http://commonworlds.net/research/.
Crutzen, P., and E. Stoermer. 2000. "The Anthropocene." *IGBP Newsletter* 41: 17–18.
Dempsey, J. 2016. *Enterprising Nature: Economics, Markets, and Finance in Global Biodiversity Politics*. England: West Sussex, Wiley-Blackwell.
van Dooren, T. 2014a. "Care: Living Lexicon for the Environmental Humanities." *Environmental Humanities* 5: 291–294.
van Dooren, T. 2014b. *Flight Ways: Life and Loss at the Edge of Extinction*. New York: Columbia University Press.
van Dooren, T. 2015. "A Day with Crows: Rarity, Nativity, and the Violent-care of Conservation." *Animal Studies Journal* 4 (2): 1–28.
Ericson, J. 2013. "Stick Insect May Hold Key to Antibiotic Resistance: Gut Microbe Resistant to Toxins, Infections it Couldn't Have Encountered Before, Researchers Say." *Medical Daily*. http://www.medicaldaily.com/stick-insect-may-hold-key-antibiotic-resistance-gut-microbe-resistant-toxins-infections-it-couldnt.
Fawcett, L. 2002. "Children's Wild Animals Stories: Questioning Interspecies Bonds." *Canadian Journal of Environmental Education* 7 (2): 125–139.
Fawcett, L. 2013. "Three Degrees of Separation: Accounting For Naturecultures in Environmental Education Research." In *International Handbook of Research on Environmental Education*, edited by R. Stevenson, M. Brody, J. Dillon and A. E. J. Wals, 409–417. New York: Routledge.
Fortino, C., H. Gerretson, L. Button, and V. Masters. 2014. "Growing up WILD: Teaching Environmental Education in Early Childhood." *International Journal of Early Childhood Environmental Education* 2 (1): 156–171. http://www.projectwild.org/documents/GUWresearchUNC2014.pdf.
Frawley, J., and I. McCalman. 2014. *Rethinking Invasion Ecologies from the Environmental Humanities*. New York: Routledge.

Gibson, K., D. B. Rose, and R. Fincher. 2015. *Manifesto for Living in the Anthropocene*. Brooklyn: Punctum.

Gillespie, K., and R.-C. Collard, eds. 2015. *Critical Animal Geographies: Politics, Intersections, and Hierarchies in a Multispecies World*. New York: Routledge.

Ginn, F. 2014. "Sticky Lives: Slugs, Detachment and More-than-human Ethics in the Garden." *Transactions of the Institute of British Geographers* 39: 532–544.

Hachey, A. C., and D. Butler. 2012. "Creatures in the Classroom: Including Insects and Small Animals in Your Preschool Gardening Curriculum." *Young Children* 67 (4): 38–42. https://www.naeyc.org/

Haraway, D. 1988. "Situated Knowledges: The Science Question in Feminism and the Privilege of Partial Perspective." *Feminist Studies* 14 (3): 575–599.

Haraway, D. 2008. *When Species Meet*. Minneapolis: University of Minnesota Press.

Haraway, D. 2010. "Staying with the Trouble: Xenoecologies of Home for Companions in the Contested Zones." *Cultural Anthropology Archives*. http://www.culanth.org/fieldsights/289-staying-with-the-trouble-xenoecologies-of-home-for-companions-in-the-contested-zones

Head, L., B. M. H. Larson, R. Hobbs, J. Atchison, N. Gill, C. Kull, and H. Rangan. 2015. "Living with Invasive Plants in the Anthropocene: The Importance of Understanding Practice and Experience." *Conservation and Society* 13 (3): 311–318.

Iowa State University. 2016. *Bug Guide: List of Non-native Arthropods in North America*. http://bugguide.net/node/view/32329

Jegatheesan, B., and H. Meadan. 2006. "Pets in the Classroom: Promoting and Enhancing the Socio-emotional Wellness of Young Children." *Young Exceptional Children Monograph* 8: 1–12.

Keeping Insects. 2016. *Annam Stick Insect*. http://www.keepinginsects.com/stick-insect/species/annam-stick-insect/

Kirksey, S. E., and S. Helmreich. 2010. "The Emergence of Multispecies Ethnography." *Cultural Anthropology* 25 (4): 545–576.

Latour, B. 2004. *Politics of Nature: How to Bring the Sciences into Democracy*. Translated and C. Porter. Cambridge, MA: Harvard University Press.

Le Grange, L. 2011. "Ubuntu, Ukama and the Healing of Nature, Self and Society." *Educational Philosophy and Theory* 44: 56–67.

Lloro-Bidart, T. 2014. "They Call Them 'Good-luck Polka Dots': Disciplining Bodies, Bird Biopower, and Human-Animal Relationships at the Aquarium of the Pacific." *Journal of Political Ecology* 21: 389–407.

Lloro-Bidart, T. 2015. "A Political Ecology of Education in/for the Anthropocene." *Environment and Society: Advances in Research* 6: 128–148. doi:10.3167/ares.2015.060108.

Locke, J. 2009. *Walking Stick Insects: The Perfect Insect Pet*. http://www.biology.ualberta.ca/locke.hp/walk_sticks.htm

Lorimer, J. 2015. *Wildlife in the Anthropocene: Conservation After Nature*. Minneapolis: University of Minnesota Press.

Louv, R. 2008. *Last Child in the Woods: Saving our Children From Nature-deficit Disorder*. 2nd ed. Chapel Hill, NC: Algonquin.

Margett, T. E., and D. C. Witherington. 2011. "The Nature of Preschoolers' Concepts of Living and Artificial Objects." *Child Development* 82 (6): 2067–2082.

Meadan, H., and B. Jegatheesan. 2010. "Classroom Pets and Young Children: Supporting Early Development." *Young Children* 65 (3): 70–77.

Melson, G. F. 2001. *Why the Wild Things are: Animals in the Lives of Children*. Cambridge, MA: Harvard University Press.

Myers, G. 2007. *The Significance of Children and Animals: Social Development and our Connections to Other Species*. West Lafayette, IN: Purdue University Press.

National Geographic. 2016. *Stick Insect*. http://animals.nationalgeographic.com/animals/bugs/stick-insect/

Nxumalo, F. 2015. "Forest Stories: Restorying Encounters with 'Natural' Places in Early Childhood Education." In *Unsettling the Colonial Places and Spaces of Early Childhood Education*, edited by V. Pacini-Ketchabaw and A. Taylor, 21–42. New York: Routledge.

Nxumalo, F. 2016. "Touching Place in Childhood Studies: Situated Encounters with a Community Garden." In *Youth Work, Early Education, and Psychology: Liminal Encounters*, edited by H. Skott-Myhre, V. Pacini-Ketchabaw and K. Skott-Myhre, 131–158. New York: Palgrave Macmillan.

Pacini-Ketchabaw, V., and F. Nxumalo. 2015. "Nature/Culture Divides in a Childcare Centre." *Environmental Humanities* 7: 151–168.

Pacini-Ketchabaw, V., F. Nxumalo, L. Kocher, E. Elliott, and A. Sanchez. 2014. *Journeys: Reconceptualizing Early Childhood Practices Through Pedagogical Narration*. Toronto: University of Toronto Press.

Pacini-Ketchabaw, V., A. Taylor, M. Blaise, and S. de Finney. 2015. "Learning How to Inherit Colonized and Ecologically Challenged Lifeworlds." *Canadian Children* 40 (2): 3–8.

Pacini-Ketchabaw, V., A. Taylor, and M. Blaise. 2016. "De-centring the Human in Multispecies Ethnographies." In *Posthuman Research Practices in Education*, edited by C. Taylor and C. Hughes, 149–167. London: Palgrave Macmillan.

Pedersen, H. 2004. "Schools, Speciesism, and Hidden Curricula: The Role of Critical Pedagogy for Human Education Futures." *Journal of Future Studies* 8 (4): 1–14.

Pedersen, H. 2010. "Is 'The Posthuman' Educable? On the Convergence of Educational Philosophy, Animal Studies, and Posthumanist Theory." *Discourse: Studies in the Cultural Politics of Education* 31 (2): 237–250.

Pelo, A. 2009. "A Pedagogy for Ecology." *Rethinking Schools* 23 (4): 30–35. http://www.reanz.org/files/3313/8248/0628/RS23.4f30Pelo.pdf.

Puig de la Bellacasa, M. 2010. "Ethical Doings in Naturecultures." *Ethics, Place, and Environment* 13: 151–169.

Puig de la Bellacasa, M. 2012. "'Nothing Comes Without its World': Thinking with Care." *The Sociological Review* 60: 197–216.

Rose, D. B. 2016. *We Need New Narratives* [video with T. van Dooren]. 'Environmental Humanities: Remaking Nature' massive online open course (MOOC). University of New South Wales. https://www.futurelearn.com/courses/remaking-nature

Rose, D., and T. van Dooren. 2017. "Encountering a More-than-human World: Ethos and the Arts of Witness." In *Routledge Companion to the Environmental Humanities*, edited by U. Heise, J. Cristensen and M. Niemann, 120–128. London: Routledge.

Ruid, A. G., and A. M. Beck. 2000. "Kids and Critters in Class Together." *The Phi Delta Kappan* 82 (4): 313–315.

Russell, J. 2017. "'Everything has to Die One Day': Children's Explorations of The Meanings of Death in Human-animal-nature Relationships." *Environmental Education Research* 23 (1): 75–90. doi:10.1080/13504622.2016.1144175.

Selly, P. B. 2014. *Connecting Children and Animals in Early Childhood*. St. Paul, MN: Red Leaf Press.

Spannring, R. 2017. "Animals in Environmental Education Research." *Environmental Education Research* 23 (1): 63–74. doi: 10.1080/13504622.2016.1188058.

Steffen, W., P. J. Crutzen, and J. R. McNeill. 2007. "The Anthropocene: Are Humans Now Overwhelming the Great Forces of Nature?" *Ambio* 36: 614–621.

Taylor, A. 2013. *Reconfiguring the Natures of Childhood*. New York: Routledge.

Taylor, A., and V. Pacini-Ketchabaw. 2015. "Learning with Children, Ants, and Worms in the Anthropocene: Towards a Common World Pedagogy of Multispecies Vulnerability." *Pedagogy, Culture & Society* 23 (4): 507–529.

Timmerman, N., and J. Ostertag. 2011. "Too Many Monkeys Jumping in Their Heads: Animal Lessons Within Young Children's Media." *Canadian Journal of Environmental Education* 16: 59–75.

Tsing, A. L. 2015. *A Feminist Approach to the Anthropocene: Earth Stalked by Man*. New York: Keynote lecture presented at Barnard College Center for Research on Women. Video file Accessed November 10, 2015. http://bcrw.barnard.edu/videos/anna-lowenhaupt-tsing-a-feminist-approach-to-the-anthropocene-earth-stalked-by-man/

Whatmore, S. 2006. "Materialist Returns: Practising Cultural Geography in and for a More-than-human World." *Cultural Geographies* 13: 600–609.

Wilson, E. O. 1984. *Biophilia*. Cambridge, MA: Harvard University Press.

Between indigenous and non-indigenous: urban/nature/child pedagogies

Margaret Somerville and Sandra Hickey

ABSTRACT

This co-authored paper offers Aboriginal and non-Aboriginal perspectives on the emergence of urban/nature/child pedagogies in a project to reclaim remnant woodlands. Set in the context of indigenous issues explored in a special edition of the journal on land based education, the paper engages critically with a claim by a group of ecologists, that as urbanisation increases globally indigenous languages and knowledges are being lost in parallel with the loss of species.[1] The paper analyses children's multimodal images and texts in the book, *Because Eco-systems Matter*, produced as an outcome of the project. In identifying possibilities for alternative storylines to those of loss and moral failure, the paper concludes that pedagogies incorporating contemporary hybrid Aboriginal forms of language and representation offer all children the possibility of re-imagining a traditional past into a contemporary present/future. In this present/future their learning and actions have the potential to name and change their worlds.

Introduction

It's a busy classroom in one of the poorest areas of western Sydney's growing urban metropolis where one small boy has had another major meltdown. He has become mute, will not speak to anyone, and is not able to participate in the class activity. Searching for a point of connection, Margaret remembers his very lifeful drawings in the book 'Because Ecosystems Matter' produced from a project about children planting endangered native plant species of the remnant Cumberland Plain. How did this child who appears so troubled, produce such detailed bright and meaningful drawings? Recalling the vitality of these drawings, Margaret says to him 'I've seen your drawings in the book about the Cumberland Plain'. The child looks up immediately and responds, 'what book?' clearly not quite sure what she means. 'The *Ecosystems Matter* book about the plants'. His face lights up, bright and engaged, becoming very present to that moment, 'oh yes', he says, 'the plants', and talks about his drawings. (Somerville, journal excerpt, 2015)

This child went on to develop a close relationship with the researcher and became highly engaged in the ongoing language mapping project, the site his meltdown. His participation in the ongoing project resulted in outcomes in his literacy learning that were not previously thought possible (Somerville, D'warte, and Sawyer 2016). The Cumberland Plain Woodland project offered a turning point in this child's learning, raising interesting questions about its urban/nature/child/pedagogies intersections. In response to the call for papers Sandra and Margaret decided to explore the context and nature of the relationship between the environmental education program of the Cumberland Plains project and the significant involvement of the Aboriginal English program within the school.

Environmental education and indigenous knowledges

A review of papers in the last five years (2013–2017) of Environmental Education demonstrates that consideration of indigenous issues in relation to environmental education is rarely addressed in the journal. A special issue on 'Land based education' offers nine substantial papers and a wealth of information and inspiration (Tuck, McKenzie, and McCoy 2014). Outside of this special issue there are only two articles that mention the intersection of indigenous concerns and environmental education, one of which notes the intersection of indigenous and urban in an article about urban spaces more generally (Derby, Piersol, and Blenkinsop 2015). The other is based on an analysis about the construction of Indigenous identity in Australian newspapers with some relevant comments about language use and stereotypes (Lloyd and Boyd 2014). The key themes that emerge from the total body of research are: *Land, country and place* as sites of indigenous ontologies and epistemologies; *Settler colonial critiques of environmental education*; and *Significance of language and naming*.

Land, country and place

Despite minor differences in terminology where *Land* is the key term in papers from the American continent (Tuck, McKenzie, and McCoy 2014) and *Country* is the terminology in Australia, there is broad general agreement across all of the articles that concepts of Land and Country embody profound ontological and epistemological differences to the often invisible underpinnings of environmental education. For indigenous people Land is understood as agentic, 'a teacher and conduit of memory that both remembers life and its loss and serves itself as a mnemonic device that triggers an ethics of relationship to land as familial, intimate, intergenerational, and instructive' (Tuck, McKenzie, and McCoy 2014, 9). Country is similarly understood 'as a vital interconnected web of social, ecological and spiritual relationships [that] epitomises the way of existing in and viewing the world that might be termed the 'relational ontology' of Indigenous Australians' (Whitehouse et al. 2014, 58). Concepts of Land and Country are largely established as oppositional to the concept of 'Place' in this special edition which is predicated on a response to critical place-based pedagogy (Gruenewald, 2003) from an indigenous perspective. The concept of Place is generally criticised because it 'fails to meaningfully address colonial legacies in education and particularly how conceptions of place have been involved in their continuance' (Paperson 2014, 16).

Settler colonialism critiques of environmental education

The understandings embodied in environmental education in settler colonial societies are seen as operating against indigenous ways of being in, and knowing the world, to the point of the erasure of an indigenous presence: 'The … representations of human/land relations in learning environments is a specific example of the way in which Indigenous epistemologies and ontologies are denied' (Bang et al. 2014, 44). In settler colonial societies processes of decolonization (a framing for the special edition) involve both an active critique of the processes of colonization and a simultaneous presencing of indigenous ontologies and epistemologies (Tuck, McKenzie, and McCoy 2014).

Language and naming

The significance of naming and language is evident across all of the articles in the special edition and is thus understood 'as an important feature of land education (Tuck, McKenzie, and McCoy 2014, 13). 'Naming' in learning is recognized as the 'site at which issues with references between Western and Indigenous epistemologies unfold' (Bang et al. 2014, 47). Researchers found they had to create their own language 'to express these concepts and what we want our kids to learn and understand as well as to help us to be able to become familiar with that language (Bang et al. 2014, 46). This reference to children is the only one in this collection of articles.

Urban intersections

The intersection of the urban with indigenous, and environmental education produces particularly new and different thinking. Three of the papers in the special edition specifically address this intersection (Bang et al. 2014; Paperson 2014; Sato, Silva, and Jaber 2014). In these papers Time and Space/place are both reconceptualised within indigenous onto-epistemolgies of urban land. The invisibility of both indigenous and more than human relations with the land is signified as of particular relevance in relation to urban spaces. In the Chicago wetlands, for example, 'generations of Indigenous nations have been in relations – ancestral, medicinal, migrational, and economic', indicating the collapsing of past and present time. Urban space is changed with 'the emergence of land and water in dynamic relationships and the life it supports' despite 'centuries of attempts of erasure, remaking, and geographical violence' (Bang et al. 2014, 38). Developing new forms of language and naming is identified as one of the key means through which this transformation is achieved.

These urban focused papers also make important contributions to thinking about the intersection of urban/nature/pedagogy, which can be equally applied to children. A variety of innovative methodological approaches offer potential pedagogies including *Storied land*: 'Storied land moves place back, between, and beyond to Native land, providing a transhistorical analysis that unroots settler maps and settler time' (Paperson 2014, 124); *Social mapping*: 'as a technique for developing a land education: to recognize and analyse identities of territorial resistance by recording the existence of various historically invisible social groups and understanding the socio-environmental conflicts they face in their lives' (Sato, Silva, and Jaber 2014, 103); and *Remaking Relatives*: 'in order to know ourselves and our ancestors better, we should remake relationships with our plant relatives (Bang et al. 2014, 46).

Languages and ecologies in western Sydney

Western Sydney (GWS) is one of the fastest growing and highly populated urban areas of Australia with over two million people in the 2011 census and growing at an increasingly rapid rate. It is highly multicultural with 41.1% of the population born overseas speaking over 100 different languages. It is also the home of the largest concentration of Aboriginal people in Australia with two thirds continuing to occupy their traditional lands and one third coming from elsewhere to settle in western Sydney. Traditionally home of the Dharug, D'harawal, and Gundungurra peoples, these languages of place are no longer spoken in everyday life although there are important initiatives in language revival. Aboriginal dialects of Standard English, known as Aboriginal English, are commonly spoken among people from the diverse Aboriginal language groups now living in the region.

Western Sydney is also the home of the Cumberland Plain Woodland, occupied by Aboriginal people for millennia. At the time of European settlement, the Cumberland Plain included 1000 square kilometres of woodlands and forests. The westward expansion of Sydney over the plain has placed enormous pressure on the woodlands and its ecological communities. Cleared and used first for agriculture and then for urban development, most of the ecological communities that originally flourished on the plain are now considered endangered. A Scientific Committee, established by the Threatened Species Conservation Act, made a Final Determination to list the Cumberland Plain Woodland as an Endangered Ecological Community in Part 3 of Schedule 1 of the Act. The following quotes from the determination illustrate the scientific understanding of an ecological community and the past, present and predicted future of the Cumberland Plains Woodland:

7. The Community, as defined by the proposal, satisfies the definition of an Ecological Community under the Act, i.e. an assemblage of species occupying a particular area.

8. Only 6% of the original extent of the community remained in 1988 (Benson, D. and Howell, J. 1990 Proc. Ecol. Soc. Aust. 16, 115–127) in the form of small and fragmented stands. Although some areas occur within conservation reserves, this in itself is not sufficient to ensure the long term conservation of the Community unless the factors threatening the integrity and survival of the Community are ameliorated.

10. In view of the substantial reduction in the area occupied by the Community, its fragmentation and the numerous threats to the integrity of the Community, the Scientific Committee is of the opinion that the Cumberland Plain Woodland is likely to become extinct in nature in New South Wales unless the factors threatening its survival cease to operate. (Office of Environment and Heritage, NSW Government 2016)

In this scientific definition of 'assemblage' the ecological community of the Cumberland Plains Woodlands includes the distinct groupings of plants that occur on the clay soils of the undulating Cumberland Plain. The list of plants is extensive, with the most commonly found trees, Grey Box Eucalypts, Forest Red Gums, Narrow-Leaved Ironbarks, and Spotted Gum, forming an upper canopy and native shrubs, grasses and herbs growing as an understory. The woodlands provide habitat for many species of small creatures including birds, reptiles, insects, small and large mammals, who make this place their home. Others, such as the emu, have disappeared from these remnant stands, leaving its only trace in the name of the suburb of Emu Plains where emus were sighted at first settlement. According to the Scientific Committee's assessment the Cumberland Plain Woodland itself is likely to become extinct unless 'the factors threatening its survival cease to operate'.

Species extinction and language loss

The separation of nature and culture in Western language, thought and practice has long been recognised as the reason for the destructive exploitation of the natural world (e.g. Plumwood 2002; Rose 2004). More recently there has been increasing recognition that has resulted in massive changes to the capacity of the Earth to sustain life with the proposal of the new geological era of the Anthropocence, the time of human entanglement in the fate of the planet (Zalasiewicz et al. 2010). In an early article that references the new epoch of the Anthropocene, ecologists address the separation of nature and culture in scientific thinking (Pretty et al. 2009). They argue that previously connections between biological and cultural diversity have been considered separately but need to be understood as interconnected. They cite a growing body of literature supports the idea that linguistic, cultural and biological diversity are interlinked and that diverse languages and knowledges are being threatened by the dual erosion of biological and cultural diversity. A close relationship between species diversity and indigenous language diversity is identified, proposing that the loss of one (local knowledge and practices) results in a concomitant loss of the other (ecological integrity) and vice versa (Pretty et al. 2009). Indigenous languages encode collective knowledge in a way that is often non-translatable, but links speakers to their landscape inextricably. The ecologists recommend that 'any hope for saving biological diversity is predicated on a concomitant effort to appreciate and protect cultural diversity' (Pretty et al. 2009, 100).

Cumberland plain woodlands project

Willmot Public School is located in the highly urbanised Mt Druitt area of western Sydney. The school's website describes a population of over 20% Aboriginal children and 40% multicultural children who come from many different countries and language groups. In 2014 Greening Australia initiated a partnership with the school to address the lack of diversity in a remnant stand of eucalypt trees in the school grounds. Like most of the remnant stands of Cumberland Plain Woodlands in western Sydney there was no native ground cover underneath the canopy of tall eucalypt trees. The Cumberland Seed Savers Project produced over 1000 plants from native grasses and wildflowers for the children to plant in their woodlands. Students in Years 3–6 planted about 30 species of near extinct wildflowers and native grasses within the 1000 square metre stand of trees in the school grounds. They learnt about the traditional uses of the plants from an Aboriginal elder and were taught how to use GoPro cameras to film a documentary of the project. They produced drawings and wrote stories which were compiled for the book, *Because Ecosystems Matter*, and presented a public exhibition about the 'past, present and future' of the Cumberland Plain Woodland.

'Because eco-systems matter'

The book begins with a brief one page introduction to the notion of ecosystem as including 'all the living organisms that exist together in an area. Plants and animals in an ecosystem interact with each other and with the non-living elements around them such as climate water or soil' (Office of Sustainability, University of Western Sydney 2013, I). This is followed by a Foreword from the NSW National Parks and Wildlife Service about the remnant Cumberland Plain which 'evolved over millions of years through floods, drought and fire' but was changed when 'the newcomers' arrived (p. II). There is then a Foreword by Greening Australia who partnered in the planting project: 'Over 1000 plants from over twenty rare and beautiful native wildflower species were grown for the students to plant in their woodland' and over several months students from Years 3 to 6 also developed visual, oral, written and musical materials' (p. III).

On the next page, *Our Planting Day at Willmot* (p. IV) two small children's stories are included that describe their perceptions of the day:

> We have lots of gum trees in our playground. Underneath them is very bare, not much grows there. We decided to grow the native bush back. We planted kangaroo grass and some native buttercups … Did you know that kangaroo grass isn't called that because kangaroos eat it? It's called that because the seed head looks like a kangaroo. (Ashely, Class 4–5H)

In this telling Ashely has taken on the decision and actions of their planting and adds her fascination with the idea of a grass named after the iconic kangaroo. Up to this point, however, the book follows a fairly predictable trajectory of environmental education involving young children in what might be regarded as place-based pedagogy in their school grounds. So far there is an implicit assumption that urban development is bad, that there has been much loss of species diversity and natural landscapes in western Sydney and that these children will learn about their environment through planting the remnant Cumberland Plain. It is clear however, that they learned so much more.

On the next page the book takes a decidedly different turn where the school principal provides a Foreword that begins with a reference to Gurindji Elder Vincent Lingari:

> His story was celebrated in the song 'From Little Things Big Things Grow' which is an iconic anthem for our nation. Heroes like Vincent have taught us to fight for what we believe in and that you don't need an army to make a difference but you do need 'ideas' people, planners and workers. This combination of power and perseverance can see little ideas grow into big realities. (Ms .Anne Denham, Principal, Willmot Public School, V)

This reference to a shared sense of nationhood through an iconic Aboriginal activist for land, and a song that celebrates his actions, marks a different trajectory through which this book shares in the ontologies and epistemologies described as characteristic of Land based education (Tuck, McKenzie, and McCoy 2014).

Aboriginal English program

The next page, the beginning of the actual description of the project, marks the move into the school's specific Aboriginal language program noting that 'Aboriginal children speak many different languages and dialects and the majority of Aboriginal children going to school today speak a form of Aboriginal English in the home'. One child's story of the Cumberland Plains is then included in both Standard Australian English and Aboriginal English:

> Years ago on the earth before buildings, there were trees around the world, and the animals lived there. I drew this because everyone needs to know how animals lived years ago. When people started building houses the plants and animals were dying, no trees, no animals, no food.

> Long ago da Cumberland Plains on da Darug peoples land Dere was tree aroun ebry where an da animals libed dere. I drawed dis caouse ebryone nees to no ow animals libed. Wen dem fullas dy noked up dem dere houses da plants for food and da animals r dyin, no tree, no animals, no food. (Office of Sustainability, University of Western Sydney 2013, VII)

The significance of language and naming is recognised in this translation, reinforcing their central significance in an Indigenous framed environmental education (Tuck, McKenzie, and McCoy 2014, 13). The Aboriginal English program was introduced by Sandra at Willmot Public School through her innovative research into the use of Aboriginal English as a dominant mode of communication for Aboriginal people in the urban communities of western Sydney.

> We were looking at 'language for purpose' and we decided to look at Aboriginal English, and the purpose for that being developed and the reasons why it was important for Aboriginal people to learn to speak English. Then I did a bit of research with a few people around Mt Druitt and the language that they used at home, which is Aboriginal English, and then I got onto Facebook. I have family spread throughout Australia and all my friends are from different Aboriginal clans, and we all speak Aboriginal English on Facebook. I was very surprised about that. And then I was able to see that the language was basically the same from up north to out Walgett, Bre, then down to Wagga Wagga, into the Riverina and down south again to Wollongong and all those places, so the language was still remaining the same.
>
> So that's where we started, the research on Facebook was the best research I could have ever done, and talking to people that are broad users of Aboriginal English, they use it on a daily basis, and then I spoke to ladies who are in the workforce, they have to change that language. So it's looking at when and where do we use Aboriginal English because I don't use it in the workforce, I can't use it at work, but I teach it in a classroom. When I go home I switch back to Aboriginal English, because my kids only understand that. When I'm cranky I'll soon tell them in Aboriginal English what I want them to do. That's when they know I'm fair dinkum.
>
> I think you have to have history as well as the English to see why it was necessary to develop this Aboriginal English, so they have to know the history behind it, to give the reasons as to why it was necessary to develop this Aboriginal English. We do a great history lesson, we do lots of Aboriginal history, years 5 and 6 I take right through from invasion right through to now. Kindergarten we actually do a lot of their history as well, but in another way; it's not as traumatic as what I'd give year 5 and 6. K-2 we do a lot of language but we talk about games that Aboriginal people would play, we talk about the hunting, food sources, clothing and shelter. Whereas with 5 and 6, I will take them into invasion and the atrocities that happened in those areas that became an important part of developing a language. (Hickey et al. 2016)

Sandra's Aboriginal English program is about much more than just language, and has much in common with Land based education as it is described in the special edition of *Environmental Education Research*. The use of Aboriginal languages is intimately tied to 'Aboriginal histories', enacting the necessary analysis of 'settler colonialism' identified as an essential component of land education (Tuck, McKenzie, and McCoy 2014). In the context of western Sydney, Aboriginal English embodies the Aboriginal diaspora, the largest population of Aboriginal people in Australia. A third of these people have come from elsewhere to settle in this first location of massive dispossession and loss of land at white settlement. These characteristics of language, land and people in western Sydney shape the hybrid inventive character of the intersection of urban/child/nature/pedagogies in this space.

In a hyper contemporary move Sandra found that Aboriginal English thrives on Facebook as an ongoing means of expressing Aboriginal cultural practices. Its use also maintains identification with particular Aboriginal languages that are linked to Country, as contemporary urban Aboriginal people are.

> I do the Aboriginal English classes one term in the school year and we focus on the language use there, and how it's the language that is used throughout Australia. It's not restricted to one area, not just from my area, it's also used in Moree, up north, down south and north-west. So it's a dialect that's used right through Aboriginal culture. It also a form of identification as well, because when you're speaking Aboriginal English you may throw in an Aboriginal word from your background, which will automatically identify where you're from, so if I say 'koobichee' and things like that they'll know I come from down south. So they're using Aboriginal English but they're also using their language as well, which is where the identification comes in. You're able to identify people from their Aboriginal English and the language that they have.

The Aboriginal English program is a contemporary cultural phenomenon that embraces both discontinuity and continuity for Aboriginal people who have come to live on the Cumberland Plain. In addition, the program is not restricted to Aboriginal children. All children at Wilmott Public School learn Aboriginal English as well as the histories and cultural stories that produced the language dialect. The children who participated in the Cumberland Plains project also learned Aboriginal visual symbols, plant use, and stories of the animals that once lived there. The ways that they combined their knowledge of Aboriginal history, language and visual signs and symbols into their experience of the Cumberland Plains project can be read in their drawings and stories.

Three children's drawings

A drawing by 'Achilles 3A' is the most vibrant of all the children's artworks. The page is filled to the very edges, framed by a blue sky and blue lines of rain above and brown earth below with green lines of grasses growing into the white space. Symbols proliferate in a full range of bright colours – blue, brown, red, yellow, green and black. Everything is mobile, with wiggly lines of snakes, arrowed bird prints, and flying birds moving and merging into each other and across the white space. The curvy lizard tracks, lines of bird prints, flying birds and butterflies inhabit the page as if it is viewed from above, a mixture of a western vertical orientation and Aboriginal horizontal imaginary. Bright red flames of a campfire and an Aboriginal flag are placed in the very centre of the page. Achilles' Cumberland Plain Woodlands is a busy vibrant place full of life, movement and energy.

Sandra says:

This is an Aboriginal child Achilles. He's always putting Aboriginal perspectives in his drawings. He's always using the flag, I mean he's an Aboriginal kid, he's quite proud of his culture and he'll talk to you about his culture and his dancing and all that sort of stuff. He'll include that sort of artwork in everything he does.

Achilles' participation in the Cumberland Plain Woodlands is continuous with his activities in the Aboriginal English program at the school. Engagement in the woodlands brings meaning and liveliness to his school learning. The drawing is rich in Aboriginal signs and symbols bringing the traditional past into an urban present. The urban present is also included in the traditional past with the contemporary marker of identity, the Aboriginal flag, featuring in his drawing of the time before white settlement. This drawing depicts a richly inhabited Aboriginal present rather than a sense of a lost past.

Another drawing by 'Jess, 3A' is also striking for its dominance of Aboriginal symbols. The white space of the page is highly ordered with neat bounded images in lead and coloured pencil. As with Achilles' drawing there is a blue sky at the top and brown earth at the bottom but most of the images are arranged along the brown earth in a more typical landscape format. There are two trees with a shelter underneath on one side, and a single tree on the other. In the centre a fire and campsite is perfectly executed in Aboriginal symbols. The fire emits smoke into the sky and a series of lizard and animals tracks also leads into the sky, disrupting the vertical orientation and any smooth reading of this drawing as distinctively Aboriginal or non-Aboriginal in style.

Sandra explains:

Jess is a non-Aboriginal kid, but because I do a lot of Aboriginal art, and a lot of Aboriginal stuff here, they tend to use it – but Jess is – she's very English.

The campfires – that represents a campfire with people, someone's leaving. Oh, no, there's another campsite I see, and that's the flame and she's done the smoke, the smoke's another symbol. That could be a river or it could be storms, that googly line she's used. A lot of them took on the Aboriginal art, it's really good.

The pedagogies of Aboriginal English, including its visual symbolic forms, engage all children in *thinking through Country* (Somerville 2013), enacting the mixed up nature of contemporary urban life. In this mixed up present, Aboriginal and non-Aboriginal children alike bring the past into the future and the plants of the Cumberland Plain Woodlands now enter representation with Aboriginal symbolic forms.

The third artwork was made by one of the less able children in collaboration with Sandra. Jayden's artwork, created from cut out coloured paper, shows a horizontal blue sky that takes up about a third of the space, another third across the middle shows a row of trees rising up from brown earth that occupies the bottom third. Symbols include a series of half circles (seated humans); parallel dotted lines (walking tracks) drawn in white on the blue; concentric circles on a blue waterhole; and on the brown earth (camp sites); animal tracks; and an elliptical symbol for the entrance to a cave dwelling. A bright green cut out snake crawls across the bottom towards an upright brown kangaroo sitting tall amidst the trees. The rich complex array of symbols is well integrated and regular, with a pleasing balance and interesting story.

Sandra comments:

> Jayden and I both worked on this. We cut out blue sky with the paper, we cut out the trees, drew the branches in. We cut out the kangaroo – that was interesting. We cut out the snake and then just made it like a collage, everything on top of each other. There's Rain. Rain in the sky, storm clouds. That symbol represents storm clouds, so that's the rain coming on. He's got lots of storm clouds all along. Jayden has a lot of issues, so yeah, we had to sit and work that out with him and once we started cutting out and I showed him how to cut out and make the picture, he went with it.

The story of how Jayden's artwork was produced gives an insight into the Aboriginal English pedagogies in action in relation to the Cumberland Plain project. Sandra uses the materials at hand in ways that are accessible for this less able child to produce an artwork that tells a story of animals, plants, humans, weather and elements in their complex entanglements. Rather than creating a story of fear and destruction, or failure and loss to achieve an outcome in his learning, a satisfying balance is achieved through this collaborative act in which the world is imagined differently.

In these three drawings the children's images and stories about the traditional past disrupt any smooth notion of a romanticised past as opposed to a despoiled urban present. The children's artworks depict their imagining of the past life of the remnant Cumberland Plain transposed into their present reality. The oral/verbal language of Aboriginal English is extended through multiple visual symbols that populate their artworks in the form of human camp sites, fires, shelters, waterholes and walking trails alongside symbols of the more than human world of plants, animal tracks, flying birds and weather. A contemporary Aboriginal flag inserts present into past in two of the drawings of this re-imagined traditional time. There are snakes, emus, birds, lizards all appearing amongst stands of trees that look exactly like the arrangement of trees in their school playground now, rows of conventional tree drawings with spaces in between.

In these drawings the past is brought alive in the present. The children have imagined a different future world where animals come to inhabit their woodlands and all children have access to Aboriginal language words, symbols and stories. This enables a hybrid intermixing of identities in place as well as visual forms and meanings through which children can navigate their diverse ethic origins in this urban space where they have come to dwell together.

Naming their (entangled) worlds

Throughout the book the children's names such as Naulu, Aku, Tanika, Calvary, Dansia, Beniah, Atoc, signify the diverse ethic origins of children displaced from their own indigenous belongings and learning new ones in this place of the Cumberland Plain. Sandra incorporates their different ethnic and language backgrounds in her Aboriginal English teaching:

> We also been looking at where the English language actually originated from and we've looked at Arabic, Latin and the old Persia and all those places where the English language actually originated from. We compared a few words in English and Arabic and got the Arabic meaning for the English word which was actually in English. We looked at the word garbage a lot, because the word garbage doesn't exist, it's a made up word, it's a created word. So the word's actually garb, which means to sift through and separate things, and the kids all took that on board really easy. We looked at – a lot of the Arabic kids actually found that pretty interesting that it was their language that these words actually [came from] – we looked at the word Mufti Day, what did mufti mean? Mufti is an Islamic lawman and we call it Mufti Day because of the comfortable clothing that he wears. So the kids now understand where Mufti Day comes from and it's not Mufti Day, it's Moofti Day, we had that cleared by Arabic kids in the classroom on how that's pronounced.

> And we pulled on a lot of the kids with other languages, on their pronunciations and their English and how their English words can also mean something different to what non-Aboriginal people understand as or Europeans understand as the meaning of the word, it is totally different for a lot of other places. Some Aboriginal English that we've found actually relates to African language. So that was really interesting when the African kids says, 'But that doesn't mean [Speaks African] girl in their language, it means 'a hair tie'. So they were quite thrilled that these words were actually coming out and they were able to relate to the words even though it meant something different. So that's basically what I've been doing [in Aboriginal English classes].

This broad exploration of the children's origins and mobilities through the intersections of language in the Aboriginal English classes facilitates the inclusion of children from a wide range of ethnic origins within this urban space. The operation of this inclusive pedagogy is further evident in the end of the section on the traditional past which finishes with a story called 'All plants have their uses' by Mendi, Eliza and Moana. It shows how the children have understood the information they were given about the Aboriginal use of native plants:

> The banksia is a useful plant. Aboriginal people were able to use them for many different and unusual purposes. The banksia could be used for a hairbrush. If you dip the banksia flowers into water it then makes the water taste sweet like cordial.

> The dianella plant is also useful. It was a good source of food. If you had a long hot walk in the desert you could chew on the bottom of the leaf to make your tongue wet. Its strong straight leaves could also be used for weaving to make baskets, bracelets and other useful things.

> Sandpaper trees have rough leaves you could scrub on tools to stop splinters in your hands when handling them.

> Paperbark from trees could be used to build huts and shelters. It could also be used for baskets and putting food on. You could even draw on it.

> Lemon myrtle is a very strong medicinal plant. You could use it to blow your nose if you had the flu. When you smell it your nose would unblock.

> (Story by Mendi, Eliza and Moana, Class 4–5H)

Through retelling the stories of Aboriginal plant use, children from many different cultural backgrounds learn to name their worlds differently. While framed within a storyline of a traditional Aboriginal past, the children's retelling of the story creates a very present and embodied relationship between human and plant within the Cumberland Plain Woodland communities, thus 'remaking their relations with plants (Bang et al. 2014). The plants are not distant scientifically known and named species, but are experienced as intertwined with their human children's bodies – their hair, sense of taste, wet on the tongue, stopping pain of splinters on hands from sharp wooden tools, providing shelter and healing for sick bodies. Plants are experienced as evoking direct sensory responses of smell, taste, touch, wet and dry, sheltering and healing materialities that merge with human senses and human bodies.

Just like the practices of 'urban rebirth' in land based education in Chicago, these children are 'remaking relations' with their plant relatives, 'in order to know themselves and their ancestors better' (Bang et al. 2014, 46). By the inclusion of contemporary Aboriginal language, cultural practices, and visual symbols, the first section of the book *Because Ecosystems Matter* reconfigures the urban/nature/child pedagogy assemblage of the ecological community of the Cumberland Plain. It enables the development of new contemporary naming and representational practices that enact an onto-epistemology of *Country* (Whitehouse et al. 2014). Through Sandra's Aboriginal English pedagogy and cultural practices, thinking through Country is transposed into an urban/nature/child pedagogy in which all children from this complex multicultural community are included in its assemblage of plants, animals and humans.

Impact of urbanisation and the Cumberland Plains today

In the other two sections of *Because ecosystems matter* students represented their learning about the impact of urban development on the Cumberland Plain today. These sections tend to lack the vitiality of the section about the traditional Aboriginal past as if they are no longer allowed to draw on the rich creative meanings it offers to their understanding of the project. They can be read, however, as a critique of settler colonial society as a prerequisite for change (Tuck, McKenzie, and McCoy 2014). Jerome's artwork, for example, depicts building waste with a small insert of a photo of a waste dump beside his depiction using paint and crushed paper.

> This is a pile of building waste. Most of it could be recycled or used again on other projects. You may think of it as rubbish but it could be something different. You should be careful when removing left over metal as it could be really sharp. The land where the rubbish is could be used for a family's home or park instead. (Jerome)

In their representation of the Cumberland Plains of urban development children have made artworks, as instructed, using simple cut out shapes of standardised houses trees to represent the absence of vitality and the wild of life. The accompanying texts commonly repeat the same story of the nature of urban development:

> Many years ago there were many trees and animals, now there are houses, factories, building and so much more. Animals now have nowhere to live. They are dying and getting run over by cars. (Rory)

These images of the present stand in stark contrast to the possibilities opened up when considering the traditional (Aboriginal) past. The line drawings in coloured pencil tend to repeat the same story with lifeless standardised houses and roads and an absence of trees and other forms of life. In terms of the necessary critique of settler colonialism in land based education they lack the rich imaginings and meanings identified previously. In reading these artworks and texts this aspect of the pedagogical approach appears to be a problematic characteristic of the general discourses of environmental education offered to these urban children in which the past is presented as an idealised state and the present as loss, leaving little option for children of a different future.

There are three representations in these sections that stand out as demonstrating an ability to disrupt the standard discourses of loss of an idealised past. 'Michael' gives agency to rocks, telling how they help the plants to 'stay up' in his story. 'Rory' has trees that remain standing because they need to breathe, and 'Nikki' simply refuses to see her place of houses and parks in negative terms.

Michael's artwork has a background of a rock in delicate shades of brown with some darker brown mossy overlays superimposed with two cut outs of two fleshy human hands each beside a small plant. The delicately painted bright green fern is the only strong colour in the image, standing out from hands, rock and moss all merging in light and dark shades of browny fawn.

> This picture shows the rocks helping the plants stay up. This plant can grow up to two metres high and if we get a lot of rain it will regrow the Cumberland Plains. I like to plant new things and watching them grow. I like that Wilmott is doing the same thing so animals can come back to the environment. (Story and Artwork by Michael Class 5–6)

Michael's hands come into existence through their activity of planting, and they are understood as part of the agency of rock that will help the plants to grow, of the vibrancy of the plant that can grow up to two metres high, and of the rain that will regrow the Cumberland Plains for the animals to return. In this way activity of planting opens up an urban/child/nature pedagogy in which Michael understands himself as inseparable from the urban/nature in which he comes into being in his mutual entanglements with plants, rock and rain.

Beniah's cut outs show a standard house with two trees, one lying down and the other upright. Beside the large image there is a smaller one that shows only the two trees.

> Today people chop trees down. Some of the trees stay up because they need to breathe. They build houses because they need shelter. (Beniah)

In this Beniah challenges the idea that trees are simply the passive recipients of urban development by depicting a tree that resists the felling and remains stridently upright in its growth because of its 'need to breathe'. This need to breathe is born of a relation between child and tree, sharing the same need, the same air, through which the tree provides life sustaining breath for the human child.

The final section past and present combined

In the final section children are immersed in their planting and imaginings where past present and future are collapsed into one. Many of the delicate and beautiful drawings in this section actually incorporate leaves from the gum trees that form the remnant Cumberland Plain in their school ground. Their planted wildflowers proliferate, animals dance, butterflies fly, birds and insects populate the air of trees. Nikki's artwork represents her desire to understand her urban life as a beautiful expression of past, present and future. Her coloured pencil line drawing is scattered with pressed leaves, each imperfectly beautiful with their veins, insect holes and subtle shades of green, grey and fawn. There is a black road along the bottom with the lines that show the cars where to go, tufts of green and purple grass, a house and a playground with a larger and smaller stick figure on a swing. A big pink and purple butterfly flutters above the house beside a flock of black birds and the sun shines down from one corner as tufts of grass climb up each side of the drawing to reach the faint line of blue sky above.

> In the daylight the Cumberland Plain is beautiful but it is only fragments across Sydney. Lots of trees, wildflowers and animals were taken away to make way for houses, schools and parks. I think it looks beautiful now and with a couple more trees and plants it will look amazing. (Nikki)

In her image and story Nikki has perfectly collapsed past present and future urban space. She under-stands how the Cumberland Plain now exists imperfectly in fragments, recognising the loss of wild-flowers and animals to provide for human life in this urban space. She is simultaneously hopeful that their activities in planting will add to the beauty she already experiences in her place for an enhanced future life of all it embraces.

Sandra's approach to the project similarly destablises chronological time and the space of urban loss. She notices the ways that children connect to the past-in-the-present, the frequent identification of all children with the Aboriginal flag, and the purpose of the project in relation to introducing 'natural vegetation and encouraging native animals back. So the kangaroo and the snake represent their coming back to where they've now done all this garden'.

> Just to rejuvenate the parts down here, and Greening Australia loved it because of the canopy, the trees, and the protection the grasses and that would have had, and that's why that area was chosen, and they called this part here the last of the woodlands. These trees are the last of it.

> We have had some lizards and snakes coming back into the school. I don't know whether it's because it's cold now and they're looking for somewhere to nest, or because of the grasses. We won't get the kangaroos because of our fences and things, but if we can get some of the native birds, we get lots of birds, we can hear them all around us.

In considering the project in the broader context of children's out of school lives and places, Sandra evokes a very different urban/nature/child assemblage in relation to the ecological community of the Cumberland Plains:

> Most of our Aboriginal kids don't ride in these streets. They head straight for that back bush, right down the back – the old air force base was down there and they've now since closed down and it's all bush down there. All of them, and I know my lad too, and his mates, I say 'Where'd youse go?' they say 'oh we went down to the red hills'. It's only a little lump, but it's red dirt. They call it the red hills, they go to this little gully type of thing – yeah, but it is all bush down there.

> They just love getting' down there, they catch lizards, they build tree houses – it took them weeks to build this tree house, every weekend he was heading down there 'oh, we gotta finish off the tree house'. 'Isn't that thing finished yet?' Yeah, so they spent quite a long time down there doing that.

For Sandra, as with the other Aboriginal parents, children's learning happens everywhere (Somerville in Woodrow et al. 2016). Aboriginal Education Officers bridge the lives of children outside of their school learning, particularly Aboriginal children and their communities. In this way their pedagogies draw on the Aboriginal diaspora and continuing cultural practices maintained in the face of the struggles that living in these communities entails. They understand that children have their own ways of enacting their multiple entanglements in everyday urban spaces and places. Through this understanding their pedagogical interventions build on the vibrancy of children's everyday lives.

Conclusion

This paper is framed within the key themes of *Land, Country and place*, *Settler colonial critiques of envi-ronmental education*, and the *Significance of language and naming* that emerged from a review of the articles published in the special edition of *Environmental Education Research* on Land based education (Tuck, McKenzie, and McCoy 2014). Special attention is paid to the articles that focus particularly on the intersection of the urban with land based education and the potential pedagogies of storied land, social mapping, and remaking relations with plant relatives that these articles offer.

The Cumberland Plain project in which children planted endangered species of native grasses and wildflowers under the remnant Cumberland Plain Woodland in their school grounds in highly urbanised western Sydney was explored by an Aboriginal and non-Aboriginal author for the intersection of urban/nature/child/pedagogies.

It was found that the environmental discourses of the Cumberland Plain and the approach to environmental education through this project incorporated a chronological approach to time with the past imagined as a glorious, idealised time of connection with nature, the present a time of coming to terms with loss and environmental destruction, and the future embroiled with moral imperatives against urban development.

Because of the school's innovative Aboriginal English and cultural history program however, the book *Because Ecosystems Matter* contains an alternative storyline that has much in common with the ontologies and epistemologies Land based education. In the urban/nature/child pedagogies observed in this book, time and urban space are conceptualised by the children who bring the past into the present, the present into the past and past/present/future time exist simultaneously.

The critique of settler colonial history represented in the children's images and stories of a barren and destructive present do not seem to engender significant learning for these children. Their representations generally lack imagination and meaning in the context of these imposed discourses. When they depart from this and enter a space where they are again invited to represent past present and future, they are once again liberated into an aesthetic wonderland where beautiful drawings are often combined with real fallen leaves from their beloved remnant Cumberland Plains Woodland.

While language is the crucial basis of their creative responses, it is the richness of the pedagogies brought forth in the Aboriginal English program that appears to underlie the extraordinary and transformative potential realised when indigenous approaches and environmental education are genuinely integrated. Its distinctive contribution is the inclusion not only of Aboriginal English, Aboriginal languages, cultures and histories, but the ethnicities of the diverse children who live together in this urban space. Significantly it also includes the more than human world, remaking relations with plants as well as providing habitat for all of the creatures that come to live in this space.

And, finally, as Sandra reminds us, children seek out wild spaces too in their everyday lives outside of school and find them in the urban spaces in which they live.

Notes

1. Lower case indigenous is used to signify indigenous people in a general sense, upper case Indigenous is used for specific Indigenous people. In most cases Aboriginal is used rather than Indigenous for Aboriginal Australians who prefer this to the term Indigenous.

Acknowledgements

We would like to acknowledge Willmot Public School, Greening Australia, and the Office of Sustainability Western Sydney University for the implementation and representation of the Cumberland Plains Woodland project. We would also like to thank the editors of this special edition for their careful reading and feedback on the paper and the anonymous reviewers, all of whom have contributed significantly to the final version.

Disclosure statement

No potential conflict of interest was reported by the authors.

References

Bang, M., L. Curley, A. Kessel, A. Marin, E. Suzukovich III, and G. Strack. 2014. "Muskrat Theories, Tobacco in the Streets, and Living Chicago as Indigenous Land." *Environmental Education Research* 20 (1): 37–55. doi:10.1080/13504622.2013.865113.

Derby, M., L. Piersol, and S. Blenkinsop. 2015. "Refusing to Settle for Pigeons and Parks: Urban Environmental Education in the Age of Neoliberalism." *Environmental Education Research* 21 (3): 378–389. doi:10.1080/13504622.2014.994166.

Gruenewald, D. 2003. "The Best of Both Worlds: A Critical Pedagogy of Place." *Educational Researcher* 32 (4): 3–12.

Hickey, S., D. Tryst, T. Lee Bell, M. Somerville, and K. Power. 2016. *Aboriginal Perspectives on Building Children's Linguistic Repertoires to Enrich Learning: A Report of the Aboriginal Education Officers.* Kingswood: Centre for Educational Research, Western Sydney University. http://www.uws.edu.au/cer/research/research_reports.

Lloyd, D., and W. E. Boyd. 2014. "An Exploration of the Role of Schema Theory and the (Non-indigenous) Construction of Indigenous Identity." *Environmental Education Research* 20 (6): 795–813. doi:10.1080/13504622.2013.833592.

Office of Environment and Heritage, NSW Government. 2016. *Cumberland Plain Woodland Endangered Ecological Community Listing. NSW Scientific Committee – Final Determination.* Accessed 20 April 2016. http://www.environment.nsw.gov.au/determinations/CumberlandPlainWoodlandEndComListing.htm.

Office of Sustainability, University of Western Sydney. 2013. *Because Ecosystems Matter. A Partnership Project with Greening Australia, Willmot Public School and the Office of Sustainability.* Kingswood: University of Western Sydney.

Paperson, La. 2014. "A Ghetto Land Pedagogy: An Antidote for Settler Environmentalism." *Environmental Education Research* 20 (1): 115–130. doi:10.1080/13504622.2013.865115.

Plumwood, V. 2002. *Environmental Culture: The Ecological Crisis of Reason.* Abingdon: Routledge.

Pretty, J., B. Adams, F. Berkes, S. Ferrriera de Athayde, N. Dudley, E. Hunn, L. Maffi, et al. 2009. "The Intersections of Biological Diversity and Cultural Diversity: Towards Integration." *Conservation and Society* 7 (2): 100–112. doi:10.4103/0972-4923.58642.

Rose, D. B. 2004. "The Ecological Humanities in Action: An Invitation." *Australian Humanities Review*: 31–32. http://www.lib.Latrobe.edu.au/AHR/archive/Issue-April-2004/rose.html.

Sato, M., R. Silva, and M. Jaber. 2014. "Between the Remnants of Colonialism and the Insurgence of Self-narrative in Constructing Participatory Social Maps: Towards a Land Education Methodology." *Environmental Education Research* 20 (1): 102–114. doi:10.1080/13504622.2013.852654.

Somerville, M. 2013. *Water in a Dry Land: Place-Learning through Art and Story. Innovative Ethnography Series.* London: Routlede.

Somerville, M., J. D'warte, and W. Sawyer. 2016. *Building on Children's Linguistic Repertoires to Enrich Learning: A Project Report for the NSW Department of Education.* Kingswood: Centre for Educational Research, Western Sydney University. http://www.uws.edu.au/cer/research/research_reports.

Tuck, E., M. McKenzie, and K. McCoy. 2014. "Land Education: Indigenous, Post-colonial, and Decolonizing Perspectives on Place and Environmental Education Research." *Environmental Education Research* 20 (1): 1–23. doi:10.1080/13504622.2013.877708.

Whitehouse, H., F. Watkin Lui, J. Sellwood, M. J. Barrett, and P. Chigeza. 2014. "Sea Country: Navigating Indigenous and Colonial Ontologies in Australian Environmental Education." *Environmental Education Research* 20 (1): 56–69. doi:10.1080/13504622.2013.852655.

Woodrow, C., M. Somerville, L. Naidoo, and K. Power. 2016. *Researching Parent Engagement: A Qualitative Field Study.* Kingswood: Centre for Educational Research, Western Sydney University. doi:10.4225/35/5715bcdd2df24. http://www.uws.edu.au/cer/research/research_reports.

Zalasiewicz, J., M. Williams, W. Steffen, and P. Crutzen. 2010. "The New World of the Anthropocene 1." *Environmental Science and Technology* 44: 2228–2231. doi:10.1021/es903118j.

Going back and beyond: children's learning through places

Claudia Díaz-Díaz

ABSTRACT

In 1919, the Canadian province of British Columbia (BC) established the Elementary Correspondence (EC) School to provide formal education for children living in rural areas with difficult access to a school. Through children's letters, this paper interrogates the concept of place, a key one for placed-based approaches to environmental education. Place-based education has focused on natural and rural environments to engage students in environmental learning, sometimes presenting the desirability of connectedness to nature and the local community with nostalgia and romanticism. Children's letters suggest that their learning has long been shaped by their places, material and imaginary, and that the EC school was a place as important as nature in children's learning and schooling. Children's letters contribute a rich historical perspective that helps to rethink the rural/urban divide into a relationship of mutual constitution of childhood and nature across time and space.

Introduction

I travel through the forest more than the other kids, and I never take them with me because they talk about some book they have read or something they have seen. I want it quiet so I can see a moose, deer or bear on my travels.[1]

This excerpt, from a letter written in 1935 by 14-year-old Amy Pelly gives us some insights into children's learning experiences through their rural places in the Canadian province of BC, earlier in the century. This letter belongs to the BCA, containing hundreds of letters written by parents and children to their EC school teachers in Victoria, BC, in the 1920s to the 1950s.[2] In this paper, I address the urban/child/nature intersection by bringing to the present the experiences of children that lived in the past of the Canadian province of BC in a time and in a place where children struggled to get an education. While children learned through their intimate contact with nature, as Amy Pelly's letter suggests, they also learned from those places beyond their localities, like the EC School located in the city of Victoria. Although many children did not have any contact with other places than theirs own, they recreated the urban worlds in their imaginations, as lively expressed in their letters, through their teachers' letters, the books they read, or somebody else's stories. Children's imagination helped them to interpret their lives and themselves as students vis à vis to their images about the city.

Place-based approaches to environmental education focus on place to teach students to be sensitive to the needs of the environment, to understand environmental problems, and promote sustainable solutions to prevent environmental damage (Smith 2002a, 2002b; Sobel 2005; Theobald 1997). In place-based education, local places serve as meaningful frameworks to 'overcome the disjuncture between school and children's lives that is found in too many classrooms' (Smith 2002b, 593). As Iris Duhn has pointed out place is a '"social, material, and discursive field that can intensify and transform living bodies"' (Duhn 2012; 103). Somerville and Green (2015) document how the observation of children's learning experiences through their places opens up an opportunity for a paradigmatic change regarding our relationships with our environments in a time of economic, social, and environmental crisis. A number of scholars and activists have critiqued how an anthropocentric perspective – in which human beings locate themselves in a dominant ontological position regarding other living and not living entities – has pushed the planet towards a point of alarming environmental damage (Derby, Piersol, and Blenkinsop 2015; Somerville and Green 2015). These authors draw our attention to consider how human beings are intertwined with non-humans – such as animals, trees, buildings and places – which exert great influence in human lives beyond their capacity to control, foresee or even conceive such a power (Common World Childhoods Research Collective 2015; Hinchliffe et al. 2005; Pacini-Ketchabaw and Taylor 2015).

Despite the role that learning through places has in place-based approaches to environmental education, Somerville and Green (2015) argue that environmental education literature has so far neglected two important perspectives: the perspectives of primary school children and the historical perspective, that is, the approach that highlights the influence that socio-historical contexts have in educational experiences. Children's letters offer a perspective usually overlooked in historical sources – the perspectives of rural children (Gleason 2009; Halfacree 2004). In this regard, children's letters constitute a valuable historical resource to scrutinize children's notions of place in the past rural BC and re-think the rural-urban divide over time and space.

For children living in rural and distant places in the province of BC accessed in the past, the EC school was the only means to get formal education. They communicated to their EC school teachers as part of their schooling experience. Through these letters – non-living agents –children were able to connect with places beyond their own localities, which fed their imagination and expanded their identities. The travelling letters had the capacity to connect, on the one side, the rural places and natural environments in which these children lived and, on the other side, the urban world that came along with the EC school. Likewise, the EC school became a place as important in children's learning as their local environments.

In this paper I show how children have long been influenced by and intertwined with their places by interrogating the very concept of place – a core one for place-based educational approaches to environmental education (Smith 2002b; Sobel 2005; Theobald 1997). Through exploring how they experienced places and related to them, I show that children's perspectives on their life worlds add different values about place and space than those of adults (Diaz-Diaz and Gleason 2016; Matthews, Limb, and Taylor 1998; Somerville and Green 2015). Also, as a conceptual framework, the interrogation of the concept of place offers the opportunity to connect and understand how distant places, like the remote and rural places of BC with the EC school located in the city of Victoria, played a role in children's lives and their schooling experiences.

I read over 1000 letters written from 1929 to 1940 by children and their families who were enrolled in the EC school of the then-named Department of Education of BC. I selected a sample of nine students' files, which contained the largest number of letters and the widest coverage of years. These files include letters of children and their parents. Most importantly, I selected these letters because they spoke not only about children's perspectives of their local places and their everyday lives in the remote rural BC, but also about their perspectives of the places beyond their locality, including the EC school. To protect children's and their families' identities, I use pseudonymous throughout this paper.

To support my argument, first, I explain briefly why it is important to interrogate the concept of place from the perspectives of children. Then I provide some background of the EC school and the constraints that past rural BC imposed to children in getting an education (e.g. no reliable mail service, isolation). Third, I discuss how children's natural places influenced their schooling and how their experience with EC school left traces in children's learning and in their relation to place. Finally, I provide some concluding thoughts about how children's stories may have important pedagogical implications for environmental education.

Interrogating the concept of place

A romantic notion of place – the one that present nature and local communities as the desirable goal for environmental education (Ruitenberg 2005) – has also given rise to the celebration of the rural over the urban. Karsten (2005) argues that the city is often associated with problems of pollution, insecurity, scarcity of play and green places, whereas the rural is associated with an idyllic and romantic image of a place deeply connected with nature. Amy Pelly's letter, quoted above, can be easily celebrated as the portrayal of an ideal childhood, innocent and embedded in nature, reinforcing the rural-urban divide based on unitary notions of natural places and childhood. However, the value of children's letters like Amy's is that they offer critical and nuanced understandings of place in the past that can be considered in contemporary place-based pedagogies by re-thinking the rural-urban divide over time (Nairn, Panelli, and McCormack 2003). Children's local natural places in the past represented complex sites, sources of both, learning and enjoyment as well as of isolation and precariousness (Diaz-Diaz and Gleason 2016). In order to get an education, children in rural BC had to negotiate their role as students with their role as workers (i.e. farmers or housekeepers) as well as a number of constraints of their rural lives (e.g. unreliable mail services because of difficult access or harshness of environment).

The celebration of the intimate connections with local natural places as the true and only way to foster a conscious environmental citizen excludes important knowledges and experiences that are beyond the 'diffuse' and 'unclear' boundaries of the local (Ruitenberg 2005). The definition of place can also fall into essentialist conceptions that assume the existence of supposedly inextricable bond with the place where we live. When this bond is disturbed, for example, due to physical relocations, it is expected that the identity of an individual can be decontextualized, leaving a negative impact on her sense of belonging (Gruenewald 2003; Orr 1992). However, Ruitenberg (2005) points out the fact that individuals' identities are formed not only by the current location they occupy, but also by their connections with all the other places that have been present in their lives.

In the same manner that past places leave traces on who we are and on how we see ourselves, the definition of the local is only possible through the interplay between what Ruitenberg (2005) calls the *hereness* and the *thereness* or the local and the trans-local. Individuals' ideas of their localities are relationally produced in connection to their ideas and experiences of the translocal. Simultaneously, what they see as translocal is produced by their ideas and experiences of their locality. This also implies that individuals' actions do not stay in the local but are rather implicated in multiples paths: 'the result and consequences of my actions are unstoppable, trans-local, and nomadic' (Ruitenberg 2005, 216). In the case of the children's letters, their trans-local imaginings of the urban shaped their understandings of themselves as learners in rural remote BC.

The EC school in the past rural BC

In 1919, the BC Department of Education was the pioneer in Canada in establishing an EC school for children living in remote areas of the province and with difficult access to schools (Dunae n.d.). Such remote places included lighthouses, lonely ranches, fishing stations, new mining districts, and logging camps.[3] Children's schooling experiences were diverse and as mentioned earlier, letters in the BCA provides only a glimpse of the diverse schooling experiences of children. For example, the time of lesson completion depended on the reality of each family. Children completed the work of one school grade in approximately one school year, although this varied according to many factors. These included student's ability, family support of their schoolwork, and unexpected family events, like sickness[4], among others. Many children had poor mail service, which delayed receiving and sending completed lessons to their teachers. Also, as detailed in this paper, the time spent by many children in working, either helping their families in the field or in household duties, sometimes prevented them from setting aside time for their lessons which was an important factor for their progress as students.

EC school did not provide the 'materiality' of a regular school, by the means of buildings, chairs and desks, recess time, or face-to-face teacher-students interactions. Nonetheless it still was a 'place' for children to get an education. In particular, the letters written by EC school teachers, parents, and children, were central in materially shaping children's schooling experiences. For children these letters enacted their elementary school through the written voices of teachers, to whom they also wrote letters with interest and even affection. In effect, through their letters children channelled their childhood stories, challenging established ideas of childhood as always playful and innocent.

By the time children wrote their letters, the Great Depression strongly hit Canada causing people's saving to vanish, prompting unemployment, plummeting wages, and generating food scarcity (Great Depression of Canada n.d.). The crisis affected the exports of wheat, which caused the fall of the net farm income from $417 million in 1929 to $109 million in 1933.[5] In BC, the markets of fish, lumber, and fruit were lower than before the crisis, even though not as hard as the rest of the provinces. Undoubtedly, between 1929 and 1939, the Great Depression caused a countrywide crisis that tarnished the lives of children and their families in the rural BC. In the isolation of the rural BC, this crisis had an effect on children's education, which is revealed through some of the children's and their parents' letters. For example, in her letter Amy's mother describes how the impact of the Great Depression was amplified in rural remote areas of BC, where insufficient social support often left families completely without assistance. A situation that was particularly acute if the male breadwinner got sick or injured, as Amy's mother's letters make very clear (Grayson 1971).

Amy's family used to live in a village with limited access to mail delivery, compromising their main connection with the rest of BC. Because of this, Amy's mother constantly asked the school to send several lessons at once to prevent unnecessary delays between lessons. Starting the winter of 1935, Mrs Pelly wrote to Amy's instructor the following:

> In reply to your notice of the children's lessons not being sent to your office. Mr. Pelly is in ill health and is not able to work. We applied for relief and received a small monthly allowance just enough for groceries and by stitching them a few of the most necessary clothing. Mr. Pelly asked to be allowed enough to pay postage and school supplies but he was refused. You will understand perhaps better how conditions are when I tell you that so far we have not received Sept. allowance and are not sure of getting that.[6]

Like Amy, children living in the past rural BC, had to sort out a number of challenges to get an education. In their stories, children reveal a particular way to talk about their places suggesting that their local natural environments as well as their translocal EC school were not mere backdrops of their lives, but rather, a force that moulded their experiences of schooling, learning and places.

Children's places: making room for the EC school

Children's letters portrayed a world where adults did not necessarily play a central role in facilitating children's learning and schooling experiences. For example, children spent a lot of time walking through the woods, observing animals, harvesting, planting, or doing their lessons while their parents were working on the land, running chores, or minding their younger siblings. Although adults did play an important role in children's lives and survival, a number of letters describe how children learnt about their natural environments on their own (Diaz-Diaz and Gleason 2016).

The letters also show that children imagined their EC school as a place with classrooms and desks, teachers and students, and wished to be there. The letters allowed this imagined EC school and its teachers to come into children's homes through school practices like readings, doing math, drawing maps, or writing essays. In order to take part in this experience of getting a formal 'urban' education, children and their families had to modify their routines, transforming children's primary roles as workers and contributors to their family economy into that of students. School lessons competed for a place in children's busy routines of planting, hunting, harvesting, exploring, playing and cooking as well as the agricultural competed with the school year.

For instance, Mary Johnsons and her siblings had to routinely spend time working along with her father which impeded her to get lessons done on time: 'I am sorry that I have missed a week at time but I have to go and meet my father every day and see to the animals which take quite a time.'[7] In another letter she writes:

At present, dad is on the road working and I have to do the socking, harrowing, and disking. Joshua, my eldest brother, does not like ranch work so he had a job on the mill and Jim, my other elder brother, has been out working for the last four or five years.[8]

This shift from worker to student may have happened in urban areas as well. However, it was still the case in rural communities that children's labour was too valuable to the well being of families to be abandoned for full time study.

By 1934, despite her farm obligations in the period of the Great Depression, Mary reported to her teacher that she had set a daily time and special location for her lessons:

I do my lessons in the mornings after milking the cow. I have made a place to keep my papers in out of a box and it is just large enough to use as a table. I made a cover out of a flour bag and my books I keep in a box on the wall.[9]

However, during spring and summer planting and harvesting time, Mary could not keep the school routine as she would have liked because she had to work out in the field to help her father get enough production of food for their family and also for trade. By 1936, Amy Pelly also had a busy time in her farm neglecting her lessons in order to support her parents in their work as farmers:

I'm very busy helping mom and dad and I have no time to do lessons, and I am a little afraid to do them because if I don't make the grade the first time, some friends or (sic) of mine, if they find it out, I will have all Giscome laughing at mother because they say it is impossible for us to make the higher grades but I'm going to try hard to show that we can.[10]

Parents and children alike knew their possibilities to get secondary and postsecondary education were sometimes limited. In one of her letters, Mrs. Pelly, Amy's mother, wrote: 'Could some form of examination be given to her (Amy) whereby she could pass on its marks? She is not intending to go to high school anyways.'[11] Interestingly, as her mother wrote this note to her instructor, Amy planned to finish her grade 8 secretly, without letting her parents know about it. Amy explained her intention to her teacher:

Don't be surprised I have decided to try to complete grade 8. This is to be did (sic) secretly I do not want ma or dad to know. I will do my lessons and send them in with Lucille and Wilfred's lessons. Mother acts as if she wished I had finished whenever she speaks of it now I am going to try and do it without anyone knowing it.[12]

While Amy's mother doubted her real possibility of finishing her elementary education, Amy prepared her lessons as an unexpected gift for her mother. Amy's experience suggests that school did matter, as much as the subsistence of the family, and that any effort to assure obtaining both, education and subsistence, was worthy.

Learning through the EC school: the role of translocal places in children's lives

Little by little, the EC school introduced school practices in children's homes. Children and their families allocated a place in their houses and imaginations for the EC school to come, despite the various obstacles and challenges that the harshness of weather, the great depression, the harvesting cycle, the lack of food and school supplies, or the isolation, imposed in their lives.

For children, places were not only those of material existence but also those that children engaged through their correspondence with their teachers. In particular, children and their families' ideas and experiences of urban places in BC, like Victoria or Vancouver, played a role in their interpretations and meaning of their local places and themselves. The letters show children and their families' interpretations of their own locality in a continuous reference to the conditions of the city. While the contrast between the rural and the urban in children's letters may be interpreted as a reification of the urban/rural divide, these letters show how the places beyond their locality formed part of their lives and thus played a role in children's perspectives of their places and of themselves.

In children's isolated places, letters also served as vehicles by which children had access to experiences in others places and stories that fed their imagination. Children did not have many opportunities to leave their rural localities for vacations or other reasons. For instance, since Mary's family came from England to Canada in September 1920, when they settled down in Knouff Lake, BC, Mary had left their home only twice. She wrote:

We came to Knouff Lake, I think, in September 1920 and I have only been away twice since I came here. That was when I was four and broke my arm and last year when I went to stay for five days with some friends at Heffby Creek and went to town on the Sat., the first time since 1922. It was also the longest car ride I ever had although it was only 14 miles.[13]

In another letter she says:

I hope you had good time in Vancouver; I'd like to go there. Kamloops is as far as I have been from here since we came out from England and I have only there twice (in fifteen years).[14]

For Mary, her isolation was not only reflected in being far away from the city but also in lacking daily interactions with other children and adults within her locality. Mary wrote:

Knouff Lake is a pretty place in summer but it is very quiet. Our nearest neighbours live half a mile away and there are no girls there. I am 16 this year and sometimes it gets a little lonely with no one near my own age around.[15]

Similarly, Will Bryson told his teacher: '[D]addy went to town twenty-six miles away and went with him for it was the first time I had ever seen the town. This town is called Smithers.'[16]

Part of children's imaginations had to do with the EC school, their teachers, and the big cities like Victoria and Vancouver. In 1935, Will Bryson wrote to his teacher:

Thank you for the picture of all the teachers in the school, and also the picture of the parliament building. I think the gardens are beautiful. I can see the flagpole and war memorial. I can also see the monument to Queen Victoria in the picture and wish I was there to see them closer. It is nice to see the buildings to which the lessons are delivered. The picture of all the teachers in the school was sure a surprise, I always used to wonder what the teachers looked like.[17]

In the case of Catherine, the EC school triggered her appetite for devouring new stories of children around the world. In most of her letters she talks about the books that she read for her lessons or for her spare time. She borrowed these books from the travelling library, a kind of itinerant library passing by rural areas of BC. Through the books, Catherine experienced the overseas realities of Norway, Spain, Italy, Haiti and Mexico. Her stories show how the books she borrowed for her EC school lessons brought new worlds into her life places. Over a month she could manage reading up to four books. Most of the letters she wrote to her teacher included analysis of the themes and characters she found in those books. The books brought to Catherine's everyday life the stories of children from other places, as she mentioned in one of her reports, of Venice or Haiti. In these stories children lived adventures with fantastic and real characters from elves to exotic animals that amused and interested Catherine in her quiet locality.

Learning from places: going back and beyond

Children's letters offer a testimony of their connections with their locality as well as with the places beyond: the EC school, other towns in BC, such as Victoria and Vancouver, and places described in books, both real and imaginary. Children's experiences with their natural environments and their EC school were in a continuous negotiation, making clear that their perspectives of their own local places were multiple and even contradictory. While for a young girl the wild backyard was a fascinating place for play and discovery, for a girl in her teen years it may have been a site of isolation. These varied interpretations of their places in nature were also in a continuous interplay with their EC school, showing how all places – those material and imaginary – left traces in children's lives and not only those with which they had a direct connection, as some have suggested (Orr 1992). Children's diverse interpretations of their local and natural environments also challenge romantic ideas of nature as places of beauty and of childhood as a time of innocence. Children's letters evidenced that although they did enjoy their natural surroundings, the places where they lived where at the same time experienced as isolated ones. The latter is reflected in that they wished to be 'there,' in an urban school where they could have the opportunity of face-to-face interactions with teachers and classmates.

Children's experiences with their local as well as with the non-local places of the EC school also problematize an uncritical celebration of rural nature over urban non-nature that accentuates deep-rooted dichotomies and impede our understanding of the roles that multiple places play in children's learning and relationships with their environments. Particularly, the observation of how the local is affected, shaped, and transformed by the non-local is key to fostering a learning that is rooted in local interests but remains in a permanent dialogue and negotiation with the translocal (Ruitenberg 2005). Appreciating the rural as well as the urban can be a powerful means to foster in students an environmental awareness through the connection, appreciation, and respect for other places beyond their locality. Moreover, children's stories help to re-think the rural and the urban, rather than a divide, as a relationship of mutual constitution.

Children's letters offer insights for pedagogies that disrupt historically entrenched dichotomous ideas of children and nature (Taylor 2013). In children's letters, place was not only their local rural communities, but also their places beyond like the EC school. Interestingly, other human beings were not as salient in children's narratives in comparison to plants, animals, weather, books, and the EC school. Their stories may draw researchers' and educators' attention to relationships of difference between the diverse agents that affected children's lives in rural BC. Paying attention to relationships of difference can also provide researchers, environmental activists and educators with the possibility to think about a relational ethics that inspires human beings to re-think our responsibility in inhabiting a world with diverse more-than-human others (Duhn 2017).

Children's stories, that reconnect BCn children with their past land and history, can also inspire educators around the world to explore their own stories of children's past relationships with place and nature. Through this paper I hope to contribute to new readings of educators' everyday pedagogical experiences by bringing in a historical perspective of children's places – the material and imaginary – in which the mediation of non-humans was central for children's learning and schooling experiences. Rural and urban are constantly reconfigured in children's letters to their EC school which gives evidence of the complex interrelations that shape their real and imagined sense of place. While some children may have experienced the beauty of nature, others perceived nature in their rural contexts as harsh and limiting in their everyday experiences. Education and imagined 'real' schools in the far away cities may have appeared as places that promised better lives to those children. 'Nature' may have been less desirable than urban education with its promise of progress. 'Nature' was to be left behind, perhaps, in search of a better future.

Notes

1. British Columbia Archives (BCA), GR-0470, British Columbia (BC) Department of Education, Elementary Correspondence (EC) School, Box No. 24, File No. 51. GR-0470. March 7 1935. Hereafter BCA, GR-0470, Box 24, File 51, March 7 1935.
2. To illustrate how rural children have learned through place in the past of BC, I draw on *The Land is My School Project* lead by Dr Mona Gleason. The project focuses on children's and families' experiences of distance schooling in past BC with the purpose of informing contemporary environmental education (Diaz-Diaz and Gleason 2016).
3. BCA, GR-0470, Box 3, File 4. (n.d.).
4. BCA, GR-0470, Box 3, File 4. 1935.
5. Retrieved from: http://www.yesnet.yk.ca/schools/projects/canadianhistory/depression/depression.html
6. BCA, GR-0470 Box 24 File 51. November 1 1935.
7. BCA GR-0470 Box 24 File 17. 1934.
8. BCA GR-0470 Box 24 File 17. 1934.
9. BCA GR-0470 Box 24 File 17. 1934.
10. BCA GR-0470 Box 24 File 51. 1936.
11. BCA GR-0470 Box 24 File 51. February 20 1937.
12. BCA GR-0470 Box 24 File 51. February 20 1937.
13. BCA GR-0470 Box 24 File 17. June 8. 1934.
14. BCA GR-0470 Box 24 File 17. December 14 1935.
15. BCA GR-0470 Box 24 File 17. June 8 1934.
16. BCA GR-0470 Box 26 File 1. November 10 1938.
17. BCA GR-0470 Box 26 File 1. January 20 1935.

Acknowledgements

I would like to thank Dr Mona Gleason, Dr Claudia Ruitenberg, and Dr. Claudia Sepulveda for their insightful comments and discussions on an earlier version of this paper and the Special Issue's reviewers for their valuable feedback.

Disclosure statement

No potential conflict of interest was reported by the author.

Funding

This work was supported by a UBC Hampton Research Grant (2013-2015).

References

Common World Childhoods Research Collective. 2015. http://commonworlds.net/.

Derby, M. W., L. Piersol, and S. Blenkinsop. 2015. "Refusing to Settle for Pigeons and Parks: Urban Environmental Education in the Age of Neoliberalism." *Environmental Education Research* 21 (3): 378–389. doi:10.1080/13504622.2014.994166.

Diaz-Diaz, C., and M. Gleason. 2016. "The Land is My School: Children, History, and the Environment in the Canadian Province of British Columbia." *Childhood* 23 (2): 272–285. doi:10.1177/0907568215603778.

Duhn, I. 2012. "Making "Place" for Ecological Sustainability in Early Childhood Education." *Environmental Education Research* 18 (1): 19–29. doi:10.1080/13504622.2011.572162.

Duhn, I. 2017. "Cosmopolitics of Place: Towards Multispecies Living in Precarious Times." In *Reimagining Sustainability Education in Precarious Times*, edited by K. Malone, S. Truong and T. Gray, 45–57. Singapore: Springer.

Dunae, P., ed. n.d. "The Homeroom: British Columbia's History of Education Web Site." http://www.mala.bc.ca/homeroom.

Gleason, M. 2009. "In Search of History's Child". *Jeunesse: Young People, Texts, Cultures* 1 (2): 125–135. doi:10.1353/jeu.2010.0018.

Grayson, L. M. 1971. *The Wretched of Canada: Letters to R. B. Bennett, 1930–1935*. Toronto: University of Toronto Press.

Great Depression of Canada (n.d.). http://www.yesnet.yk.ca/schools/projects/canadianhistory/depression/depression.html.

Gruenewald, D. 2003. "The Best of Both Worlds: A Critical Pedagogy of Place." *Educational Researcher* 32 (4): 3–12.

Halfacree, K. 2004. "Introduction: Turning Neglect into Engagement Within Rural Geographies of Childhood and Youth." *Children's Geographies* 2 (1): 5–11. doi:10.1080/1473328032000168723.

Hinchliffe, S., M. B. Kearnes, M. Degen, and S. Whatmore. 2005. "Urban Wild Things: A Cosmopolitical Experiment." *Environment and Planning D: Society and Space* 23 (5): 643–658. doi:10.1068/d351t.

Karsten, L. 2005. "It All Used to Be Better? Different Generations on Continuity and Change in Urban Children's Daily Use of Space." *Children's Geographies* 3 (3): 275–290. doi:10.1080/14733280500352912.

Matthews, H., M. Limb, and M. Taylor. 1998. "The Geography of Children: Some Ethical and Methodological Considerations for Project and Dissertation Work." *Journal of Geography in Higher Education* 22 (3): 311–324. doi:10.1080/03098269885723.

Nairn, K., R. Panelli, and J. McCormack. 2003. "Destabilizing Dualisms: Young People's Experiences of Rural and Urban Environments." *Childhood* 10 (1): 9–42.

Orr, D. W. 1992. *Ecological Literacy: Education and the Transition to a Postmodern World*. Albany: State University of New York Press.

Pacini-Ketchabaw, V., and A. Taylor, eds. 2015. *Unsettling the Colonial Places and Spaces of Early Childhood Education*. Abingdon: Routledge.

Ruitenberg, C. 2005. "Deconstructing the Experience of the Local: Toward a Radical Pedagogy of Place." *Philosophy of Education*: 212–220.

Smith, G. 2002a. "Going Local." *Educational Leadership* 60 (1): 30–33.

Smith, G. 2002b. "Place-based Education: Learning to Be Where We are." *The Phi Delta Kappan* 83 (8): 584–594. doi:10.1177/003172170208300806.

Sobel, D. 2005. *Place-based Education: Connecting Classrooms & Communities*. 2nd ed. Great Barrington, MA: Orion Society.

Somerville, M., and M. Green. 2015. *Children, Place and Sustainability*. London: Palgrave Macmillan.

Taylor, A. 2013. *Reconfiguring the Natures of Childhood*. London: Routledge.

Theobald, P. 1997. *Teaching the Commons: Place, Pride, and the Renewal of Community*. Boulder, Colo: Westview Press.

Learning from cities: a cautionary note about urban/childhood/ nature entanglements

John Morgan

ABSTRACT

This article examines the urban and pedagogical imaginations that underpin the Editors' call for papers in this special issue of EER. It raises two concerns. The first is that the view of the 'urban' that underpins work in this field, whilst offering some new insights, tends to overlook the powerful forces and structures that produce urban space. The second is that in suggesting a pedagogy that is framed in terms of the 'planetary scale' and vitality of 'life itself', the processes that shape children's nature in urban spaces are obscured. The article concludes with a note about how cities are playing their part in the latest round of economic restructuring.

The issue [*of Environmental Education Research*] intends to challenge the assumption that cities are places of human and technological dominance over nature, and that childhoods are increasingly lived in human-centered unnatured urban environments. We invite perspectives that engage with urban/child/nature intersections in unexpected ways, to critically explore the materiality and diversity of nature/urban/childhoods. We want to consider what new theoretical and methodological perspectives may be useful in explorations of human/nature and nature/culture urban entanglements and allow more inclusive means for critical engagement with the naturehuman collective (Hinchliffe et al. 2005), therefore moving away from cultural universalisms about the natured child.

Purpose

This article examines the urban and pedagogical imaginations that underpin the Editors' call for papers in this special issue of EER. It raises two concerns. The first is that the view of the 'urban' that underpins work in this field, whilst offering some new insights, tends to overlook the powerful forces and structures that produce urban space. The second is that in suggesting a pedagogy that is framed in terms of the 'planetary scale' and vitality of 'life itself', the processes that shape children's nature in urban spaces are obscured. The article concludes with a note about how cities are playing their part in the latest round of economic restructuring.

The article explores literature at the intersection of three fields: urban political ecology, childhood studies, and society-nature relations. The literature in each of these is vast, so the approach has been to focus on papers and chapters that explore the educational implications of children/nature/urban relationships. The article is divided into two parts. The first part (made up of two sections) provides brief introductions to the terms 'childhood', 'natures' and 'urban' followed by a summary of the emerging 'hybrid' field of urban/childhood/natures. The second part (also of two sections) offers a critique of work in the field, focusing first on how the 'urban' is conceptualised, and second, on the limitations of the pedagogy that is proposed.

Childhood ... nature ... urban

Childhood

The last two decades have seen widespread interest in the study of childhood (e.g. Hendrick 2003; Prout 2005). In introducing the field of 'childhood studies', it is common to cite 'landmark' publications which served to challenge previously dominant notions of childhood across a range of disciplines. These include: in history, Aries (1967); in psychology, Henriques et al. (1984) and Burman (1994); in sociology and James, Jenks, and Prout 1998; in media and cultural studies, Buckingham's (2000) and Messenger-Davies (2010); and in education the edited series *Contesting Early Childhood* (Moss et al. 2016). Special mention should also be made of Colin Ward's (1978) *The Child in the City*, which added to a popular understanding of the different ways of imagining and acting as children.

These paved the way for what was termed the 'new social studies of childhood' literature which stresses the 'dynamic, social, structural, relational and interpretive dimensions of childhood' (Wyness 2006, 1). An important feature of work in the 'new social studies of children' is the insistence that childhood is a social construction, recognises the varied nature of children's lives, and attends to questions of children's agency and voice.

Nature

With their origins in the late nineteenth century, the social sciences emerged in the context of industrialisation and urbanisation. Until the last quarter of the twentieth century they operated on the assumption that 'nature' is the backdrop to the more important operation of 'society'.

Recent moves to 'bring nature back in' amounting to what Castree (2005) terms the 'de-naturalisation' of nature as practitioners in the social sciences and humanities attempt to 'make sense of nature'. In terms of the theme of this article, this process of 'de-naturalisation' extends to work in early childhood education and children's geographies which challenges the idea of the 'natural' child who is, it is claimed, closer to and more 'in tune' with the natural world. A significant contribution is that of Taylor (2013a) who, in writing as a more-than-human geographer with experience conducting research with young children, has sought to unsettle ideas about the 'natural child' through a series of studies with young people engaging with nature. Taylor's aim is to 'queer the relationship between childhood and nature' (xiii), which means to challenge the romanticism often attached to childhood and nature, in order to politicize it and place children in a more complex and situated relationship with nature. The next 'frontier' in contested childhood studies – ripe for colonisation, so to speak – is the intersection between urban nature and urban childhood.

Urban

Urban studies (based on sociological and later geographical approaches) originated in the United States in the early twentieth century as a response to the growth of the 'shock cities' of modernity. The Chicago School of Sociology described and analysed a distinctive 'urban ecology' as cities were subjected to processes of 'invasion' and 'succession' by waves of immigrants, the occupying of 'niches' and criminal underworlds (Park and Burgess 1925/1967).

As urban studies developed throughout the twentieth century, it drew upon a range of scientific, interpretive and structuralist approaches. However, until recently, such studies assumed a separation between the natural and social. As Short (2006) concluded:

> Urban theorizing has for a long time been conducted on a flat, featureless plain. For too long urban studies have ignored the physical nature of cities, the emphasis having been on the social rather than the ecological. And yet cities are ecological systems which are predicated upon the physical world mediated through the complex prism of social and economic power. (177)

This is no longer the case. The past two decades have seen the growing interest in urban nature (Wolch 1996, 2008; Benton-Short and Short 2009). Important studies that focus on the role of nature in cities included: Cronon (1991); Gandy (2002); Klingle (2007); Kaika (2005); and Davis (2003).

The result of this scholarship has been the growth of the field of 'urban political ecology' (a termed coined by Swyngedouw in 1996). Harvey (1996) noted that it is odd to find environmentalists 'excluding the massive transformations of urbanization from their purview while insisting in principle that in an ecological world everything relates to everything else' (186), adding that the long history of urbanization is one of the most significant of all the processes of environmental modification that have occurred throughout most recent world history. Heynen's (2014) review of the field charts how the 'first wave' of urban political ecology has been enriched by perspectives from feminist and post-structuralist studies, but that the common ground is that nature is a social construction and that urban natures are 'profoundly material involving the ordering, circulation and manipulation of things like food, building materials and electricity' (Latham et al. 2009, 57). However, there are important divisions in the field. On the one hand is a Marxist inspired urban political ecology which insists that there are powerful forces that shape the production of nature in cities and that cities are sites for the conversion of first nature into second nature (an approach is represented in Heynen, Kaika, and Swyngedouw 2006). On the other hand, a second approach seeks to build upon this work but draws attention to other forms of life, emphasising urban natures' hybrid qualities in which urban space is a co-fabrication between human and non-human agents and highlighting the presence and role of animals and plants in urban space, as well as feral species and domesticated animals in zoos (Wolch and Emel, 1998, Whatmore 2002; Hinchliffe 2007).

Bringing it all together ... urban/nature/children

Having discussed separately the terms 'childhood' 'nature' and 'urban', this section brings them together to form a new hybrid – 'childhood/urban/nature'. In an important statement, Shillington and Murnaghan (2016) note that, so far at least, 'children and their natures are all but absent in urban political ecology' (4), though they note that there are exceptions to the rule, such as Heynen's (2009) study of breakfast programmes in schools in Milwaukee in the United States, and Robbins' studies of the political ecology of lawns (2007), which include children in passing (more of which a little later).

Shillington and Murnaghan seek to correct this absence of children and their natures so that urban political ecology can expose the social, economic, ecological and political processes that work together to produce children's worlds. Whilst approving of how the urban political ecology literature's emphasis on flows and networks has helped to draw attention to the hybridity of the city, thereby revealing the metabolisms that shape city life and stressing the role of power relations in this process, they are concerned that 'the majority of analyses have focused far more on grand displays of power to the neglect of more intimate places and spaces' (4), whilst the focus on policy-makers not policy takers has 'skewed towards the big picture instead of fine-grained analyses of the daily lives of marginalized people' (4).

This, they argue, is where children's geographies (as part of childhood studies) can offer a necessary correction, since, from its early work in the mid-1990s, children's geographies have paid attention to the voices and agency of children and young people, and have developed innovative methods and protocols for researching with/for children. Shillington and Murnaghan's favoured approach is to endeavour to 'queer' urban childhood natures so as to reveal their 'constructedness' and develop the possibilities of alternative orderings of urban space. Queer theory, we are told, 'focuses on non-normative categories, and the destabilization of existing categorizations' (7). Whilst originally concerned with sexualities, queer theory has been extended to 'ecologies', so that seemingly common-sense assumptions about what 'urban natures' are, and notions of how 'children' are assumed to relate to 'nature' are questioned and problematized. In this vein, Shillington and Murnaghan discuss Taylor's (2013a) aforementioned work as an example of an approach that seeks to 'queer' children's nature because it interrupts commonly held ideas about the 'romantic' child existing in closer and harmonious relationship to the natural world.

Toward the end of their paper, Shillington and Murnaghan suggest ways in which their agenda might be developed, focusing on examining how children's play in nature is gendered. For instance in the ways girls play in the sand pile on the lawn, or in trees? Or, how the gendering of play leads to different ways of relating to nature? They suggest using media analysis to examine how dominant ways of thinking and acting towards nature are constructed, and propose that researchers should scrutinize how the spaces and natures produced for children in cities rely on gendered, racialized and heteronormative ideas of human-nature relations. Other examples include studies of how children negotiate their relationships with undesired urban ecologies (e.g. bugs, rats and so on) in order to explore the cultural meanings that adult society attaches to non-humans and how the boundaries between the human and non-human are produced, policed and, on occasion, transgressed.

Shillington and Murnaghan's approach to children, nature and the city –with its intent to 'queer' urban space – has much in common with that found in papers and chapters written by Taylor (2013b, 2014), Somerville (2013, 2014) and Pacini-Ketchabaw (2013), Pacini-Ketchabaw and Taylor (2015) which underpins the Editors' hopes for this special collection. That being the case, the next two sections, use examples from this emerging field to critically examine the assumptions that underpin this approach.

Exploring the urban imagination

The urban geographer Susan Smith (1994) argued that 'the interpretation of urbanism is essentially a political rather than an ontological question' (245). The study of 'the urban', she suggested, is 'as much a contest of ideas as a quest for reality; as much a statement of how things ought to be as an account of how they are' (245).

So what 'politics' informs work in the field of 'childhood/urban/natures'? The papers considered here urge educators to take an interest in the 'wild' spaces of cities. Rather than seek to take an overview of urban space, we are to take an 'under view'. The focus is on the everyday or quotidian, the unremarkable, almost as if we are concerned with 'the secret life of cities'.

This is part of a wider intellectual shift in how urban studies' imagines the city: a move away from grand narratives or over-arching models of urban processes towards a more grounded, local and partial approach. For instance, Pile and Thrift's (2000) *City A-Z* takes the form of a compendium of short essays that focus on objects within the city, such as pigeons or buses, signposts. The editors make no attempt in these essays to provide an over-arching commentary or narrative. Instead, Pile and Thrift suggest that there are different ways in which to read the A-Z or guidebook. For instance, one approach is to draw diagrams or models of the city – representations that aim to explain the social processes that create it, a second is to construct a montage which can offer us 'glimpses' or 'fragments' of life in cities, and a third possibility is to approach the city as if we were detectives, searching for clues about its meanings and motives. All of this is predicated on the fact that we can never really claim to *know* the city; at best, we can gain only partial access to it and that there are always other stories to tell.

By focusing on objects and flows of materials such an approach it is possible 'avoid the traps of the humanist ontology' which assumes that humans are the only actors that do anything of significance (Franklin 2017). Urban space is thus made up of complex entanglements of nature, technologies, flows of energy, human and non-human objects, a way of thinking inspired by social theorists Bruno Latour and Donna Haraway. It gives rise to an Object-Oriented Ontology (OOO) which suggests that we should try to 'follow the thing', be it tree stumps, tree hollows, sand, dirt, kangaroos, holes in the ground, or fences.

If this is an accurate assessment of the approach to the urban that informs work under the heading urban/children/natures, in the remainder of this section I want to focus, using examples, on three problems with this approach. These are: a 'flat ontology' of scale; the loss of focus on power; and the lack of attention to the specificities of how actual urban space is produced.

'Scale' has assumed some importance in debates about space and social theory (Taylor, 1981; Smith 1984; 1993; Marston, Jones, and Woodward 2005), following Neil Smith's (1984) argument that scale is a social production and that struggles over scale reflect unequal relations of power. A contemporary example of how scale can be enlisted in political struggle may be observed in the global financial crisis of 2007–2008 and the subsequent development of policies of 'austerity' in many large cities in the advanced economies (Detroit in the United States being the prime example). The global financial crisis was one in which the flows of capital at a global scale – the cause of the crisis – were then refracted through the scale of the urban or national, as national governments and urban administrations .were urged to take on and resolve the debt crisis that resulted from the bail-out of banks deemed 'too big to fail'. This is a classic example of the politics of scale, as power resides in being able to decide and impose the scale at which the crisis is to be resolved. Put simply, a problem that occurred at the global scale was 'outsourced' to people and places at the 'local' or 'regional' scale.

A 'flat ontology' of scale resists the idea that any one scale has greater power or influence. Thus the 'local' has its own autonomy and agency, just as processes operating at the national or global scale: a child playing in an inner-city playground has as much agency as a global corporation deciding whether or not set up its head-office in the city's downtown Somerville's (2013) discussion of La Trobe valley in the Australian state of Victoria is a good example. Somerville starts by telling us that the La Trobe valley in the heart of Gippsland is 'now the location of brown coal-fired power stations which provide electricity for the state of Victoria' (407). Electricity travels on massive overhead power lines from Gippsland to the state capital city Melbourne, 160 kilometres away. This, she says, has made her more aware of 'escalating planetary problems of global warming' (407). The fact that Gippsland is a pathologised place – for being 'poor' – is exactly what allows Melbourne to present itself as a 'clean and cultured city of the arts' (407). Both places are fundamentally connected by the lines of power (literal and figurative) that travel across the state, and Somerville offers this as an example of what geographer Doreen Massey called the 'thrown togetherness of place'. Children in this area 'negotiate this global sense of place as subject of the multiple forces of globalisation' (408). The processes of economic globalisation have, through privatisation and automation of the power industry, led to the restructuring of the economy, the loss of jobs and the depression of the local economy. Somerville's exemplary work with children in the primary school involves the drawing and discussion of place-learning maps based on their work in a reclaimed wetland near the school, a reclamation project that has been funded by one of the power companies that operates in the area.

This example illustrates the problems associated with a 'flat ontology' that places objects, people, and nature on an equal footing. The politics of scale is evident here: the processes by which the La Trobe valley became a source of energy for Victoria is an example of activity at the scale of nation as part of a modernizing project, and the economic restructuring that has led to job losses and regional decline can be seen as a product of economic forces operating at the global scale. The wetlands in which the children and teachers produce the work described by Somerville are reclaimed from former industrial land, and are paid for by a corporation that is seeking to 'give something back' to the communities that have provided its labour. The question is at what scale we should focus the analysis, and it is significant that Somerville opts in her work to emphasise the two scales of the 'local' and the 'planetary'. In the case of the place-maps which children draw, there are drawings of birds, insects and local landscape features, and children draw things that are personal to themselves and their community. What such maps do not capture is any sense of the larger political-economic forces which have shaped the place and region. This would be possible if the focus of analysis (and pedagogy) was on the power corporations and economic arrangements of which Gippsland is a part, but even here, Somerville interprets this as evidence of the escalating 'planetary' problem of global warming: in other words, a problem for a broad and undifferentiated 'humanity' rather than an issue that has been 'caused' by identifiable actors.

What is missing in Somerville's example is an engagement with questions of political-economic power, something also found in how Shillington and Murnaghan's (2016) reference Robbins (2007) work on lawns. Picking up on Robbins' brief discussion of the potential impact of lawn chemicals on children's health (103–104), Shillington and Murnaghan suggest extending the focus to consider the gender relations involved in children playing and making use of lawns, and drawing upon the rich seam of media representations of lawns and gardens. They offer some lines of inquiry, such as looking at how lawns are represented in films and popular culture, and how boys and girls use lawns for different purposes. This is certainly in line with Robbins' notion of 'lawn people', or, as his book is subtitled, 'how grasses, weeds, and chemicals make us who we are'. Robbins' point is that this complex 'entanglement' means that materials (chemicals) and non-humans (grasses, weeds) are active agents in producing certain types of subjects (people who attend to their lawns). Robbins' argument is that economic interests have a powerful effect in shaping a particular landscape type in ways that encourage ecological uniformity.

Indeed, Robbins is careful to weigh up the potential limits of thinking in terms of co-ecological production and entanglement. Part of his narrative has to do with the guilt and anxiety he experiences over the decision of whether to apply chemicals to his lawn in Columbus, Ohio. His neighbour warns him that if he does not, he will 'ruin' the lawn. And, indeed, that is what happens as the lawn is overrun by 'weeds'. Robbins at least feels good that he didn't apply chemicals, but still feels he may have 'let the lawn down'. Here is an example of the moral force of landscape and nature. And, applying it more generally, such issues are ubiquitous: 'beyond the lawn the landscape is filled to the horizon with co-inhabitants of our ecological metropolis, all 'working on us' in different ways' (137). However, all these co-inhabitants can be moved, removed, broken-down, or rebuilt. They have agency, but it is limited: the lawn can be destroyed in little more than an hour! The point is that power is not evenly shared, and the lawn only exists because of the part it plays in the process of suburban expansion that accompanied post-war economic and demographic growth in the US.

This suggests the need for an account of how urban natures are socially produced. For example, consider Shillington's (2015) article 'Birds are for girls?' in which she relays the story of the way in which her young daughter 'adopted' the birds that nested on the ivy that grew on the side of their rented accommodation in Montreal. The landlord's son is delighted when one day his father decides 'he has had enough' of the ivy and begins to clear it, prompting the boy to tell the birds to go and find a new home. These different responses lead Shillington to enquire into the gendered responses towards the birds, and she goes on to propose that these may be related to different forms of media consumption (the young girl is a fan of *My Little Pony* whilst the boy enjoys *Teenage Mutant Ninja Turtles)*. Now, whatever one makes of this story and the assumption that television viewing is so easily understood as having effects on children's identity, if this is to be taken as a way of thinking about children's urban political ecology it is one that says little about the production of actual urban nature – in this case suburban nature.

It would help to have some more detail about the suburban nature Shillington is describing. As a start we might differentiate between different types of suburban natures. Thus, the inner suburbs of cities are often rich habitats with a combination of mown grass, open grown trees and patches of flowers, shrubs and vegetables. This 'intra-urban savanna' receives artificial increments of water and nutrients as people cultivate their gardens. Insects exploit the variety of the garden, being both food for the birds but pests for the gardener, and, like weeds, they are the target of chemical sprays. In mature suburban areas trees may have grown to a substantial height and shade large parts of gardens for much of the day; they act as a noise barrier and green corridor. By contrast, the newer suburbs of cities built on former farmland have a smaller number and lower height of trees and a greater proportion of mown grass. In such suburbs the total biomass is much lower and there is a less abundant and varied fauna.

This 'suburban nature' exists within a broader economic and cultural context. Often, developers favour building larger houses at lower densities, thus requiring more land and increasing the effects on ecosystems. Such development is too diffuse to support public transport or easy walking, and therefore encourages a reliance on private auto transport which in turn relies on fossil fuel consumption.

The decision of the landlord to get rid of the ivy is presumably not simply because he 'has had enough': it is likely to be wrapped up with wider evaluations of potential rent yields, the cost of repair and maintenance and so on, which themselves take place in relation to wider discourses about the housing market and economic prospects, and cultural assessments of what is to be considered a good rental property (something linked to the different meanings and forms of suburbs in different places). In short, what is missing in Shillington's account is any sense of the specific human and non-human processes that lead to the production of a particular form of urban nature.

The examples discussed in this section explore some of the ways in which the 'urban' has, to date, been conceptualised in the emerging literature in the field of urban/children/nature. I have suggested that the assumption of a particular politics of human and non-human relations tends to favour notions of the encounter, of entanglement, of a flattening of scale, of a dispersed view of power relations, and an undifferentiated view of urban space. Despite the interest and value of the work discussed in these examples, I am concerned that it pays insufficient attention to the role of capital in the making and remaking of urban space, and thus children's urban natures. This problem becomes critical in any attempt to define a pedagogical approach.

What type of pedagogy?

So, what are the pedagogical implications of this work? A flavour of the type of teaching and learning that is envisaged can be seen from these accounts. Thus, Somerville's (2013) study of children's place-learning maps was based on a series of teaching activities undertaken in the wetlands by the local school and university teacher training faculty. The place-learning maps were a mixture of children's 'wonderings' as they explored the wetlands and an attempt to explain what they had learned through the activities. Taylor (2014) draws upon perspectives from animal studies and animal geographies to explore examples of how kangaroos were represented in children's literature at a key moment in Australian nation-building, suggesting that children were encouraged to identify with 'cute native bushland creatures' in order to cement their identification with the land as opposed to former British belonging. Pacini-Ketchabaw (2013) explores the complex relationships between children and forests. In doing so she draws attention to the politics of naming species (as 'Western' scientific names co-exist with 'folk' or indigenous names), and how a common practice of helping children identify species represents only one possible set of ties; the walk she describes involves a child being pricked by a blackberry bramble and the challenge children encounter in walking on wet moss. Her account points to the rich entanglements of children and nature.

In the final chapter of *Reconfiguring the natures of childhood*, Taylor provides a general statement of the pedagogical approach involved in studies such as these, which draw upon a 'more-than-human' perspectives in order to disrupt ideas about children and nature. She restates her argument that just as it is impossible to neatly separate nature from culture, so it is impossible to separate children's lives from the worlds in which they live with a host of others, both human and more-than-human. She proposes the need for 'common worlds pedagogies' which: (1) focus on relations of difference; (2) involve a relational ethics; and (3) understand place as a 'lively assembly of human and more than human others' (123). In addition, such a pedagogy would make use of educational enquiry as adults and children 'learn with a whole host of others' about 'worlds in which they are already located and embroiled' (123). This would include all members of the 'common worlds culture collective' and would trace threads of connection, be curious about differences, and work on the challenges and opportunities thrown up by these relations of difference. Pacini-Ketchabaw and Taylor (2015) further elaborate in their study of child-bear relations in British Columbia and child-kangaroo encounters in Australia. Their approach is to highlight that children and animals co-inhabit space in ways which 'unsettle' dominant forms of pedagogy which assume that we live in a human world that just happens to be inhabited by non-humans. The encounters they describe for children focus on 'the ethics of living together with difference' as they co-exist in worlds marked by 'throwntogetherness'.

Whilst the papers and chapters cited here are characterised by a focus on the small-scale and the 'embodied' relations of children to nature, they also contain an interesting shift of scale which makes claims about the wider significance of more than human ecologies. This is evident, for example, in Taylor's (2013a) comment that 'the specificities of their [children's] lives are shaped by the idiosyncratic geo-historical (earthly and cultural) convergences of these worlds' (115). For Pacini-Ketchabaw and Taylor (2015), such a common worlds pedagogy is motivated by the 'intensifying ecological challenges we face' (44), especially accelerating climate change and species loss as examples of 'bio-geological systems changes' (44). Citing scholars who talk about the 'Anthropocene', the authors suggest that these have 'ethical' implications for early childhood pedagogies:

> we can no longer afford the illusion of our separation from the rest of the natural world, and so educators and young children must rethink understandings of our responsibilities to the common world we share with other living beings. (45)

Similarly, Somerville (2013) ends her account of the children's place-learning maps with the words that 'place learning becomes available as a nature/culture practice for pedagogical work and opens the possibilities for new ways to teach and learn for planetary sustainability' (416). Given the terms in which pedagogies are described in these papers, it is possible to see why the editors of this special edition of *EER* point us towards Hinchliffe et al.'s (2005) paper 'Urban Wild Things: a cosmopolitical experiment'. The very first line of the abstract throws out the challenge to our thinking: 'Cities are inhabited by all manner of things and made up of all manner of practices, many of which are unnoticed by urban politics and disregarded by science' (643). These 'all manner of things' turn out to be the non-human. The focus is in the small spaces of cities (e.g. the former industrial wasteland) – and a search for new ways to engage in politics. Space is a lively, open, non-determined thing that is constantly in the making. The focus is on how cities can be inhabited with and against the grain of expert design, how urban inhabitants are heterogeneous, made up of multiple differences mobilised through human and non-human becomings, and how new politics of conviviality can be enacted. This cosmopolitics seeks to stage adventures and encounters with nature, produce new and potentially surprising wildscapes, and make space for new constellations of 'lay voices' and citizens' knowledges.

The problem with this accounts is the tendency for these pedagogical statements to be expressed in terms of some 'ex-orbitant globality' or the 'planetary scale'. The effects of this 'scalar' shift is that it serves to situate the 'real' forces that shape the world beyond reach. This is reminiscent of Nigel Clark's (2011) argument that if we extend our global historical timescales then the idea that nature is socially produced tends to fade. Viewed as such, *all* human socio-ecological relations (such as, for example, the fact that 52 percent of the planet's population now lives in settlements defined as 'urban') appear miniscule and insignificant in relation to an all-encompassing nature operating across 'deep-time'. In the long-run, humanity is at the whim of large scale environmental forces (such as tectonic shifts, volcanoes and earthquakes) that reveal the 'raw physicality' of the Earth. However, this does not mean that politics is absence in these accounts, but that the politics tends to be expressed at the local and community scales. Thus, there is an ethics of care, attention to future generations and an injunction to approach all relationships (with both the human and more-than-human) in the spirt of hospitality and co-belonging, which seems to capture the essence of what is meant by 'common worlds pedagogies'.

There is an undoubted appeal in an approach that emphasises the contingent, indeterminate and ambivalent aspects of 'entangled worlds', not least because it resonates with idea that children in both formal and informal educational settings are in the process of 'becoming'. However, focusing attention at the scales of the 'planetary' or the 'embodied' means we risk an analysis of the regional, national and global projects that shape actually existing cities. This is discussed in the concluding section.

Learning about cities

What conclusions can be drawn from this 'cautionary note'? I hope it is clear that any approach that is concerned to critically examine the processes that serve to construct nature in urban spaces is to be welcomed. The general move to 'deconstruct' categories that are too often taken-for-granted is positive, as is the recognition of the non-human and more-than-human components of urban space. It is hoped that this article may suggest how a sharper conception of the 'urban' might inform future work in the field. As a whole, the work discussed in this article serves to challenge, exhort, and motivate those of us who (as educators) have been trained, and remain trapped in, a 'humanist' paradigm (Lloro-Bidart 2015).

However, could it be that such a posthumanist political ecology, whilst useful in overcoming some of the problems associated with the 'grand narratives' of neo-Marxist political ecology and Western Science's reliance on notions of modernization, 'Edenic ecologies' and 'climax equilibrium', has tended to overemphasise 'flow', 'movement' and 'boundary-crossing' and at the expense of notions of power, borders and structure? Whilst I would not go as far as to suggest that this work is complicit with a neo-liberal capitalism which too seeks to celebrate 'mobility', 'flows' and 'border-crossing' (for a reflection on this issue, see Braun 2015), it is certainly appropriate to ask why it is that, just as the urban is being used to solve capital's crisis of accumulation and cities are being re-wired into the flows of global capital formation, 'critical educators' are re-scaling their pedagogy downwards to focus on 'lawns', 'tree-stumps', child-animal relations, or upward so that 'earthly' economic, political and environmental problems are located at a 'planetary' scale.

It is pertinent to pose this question since it should not escape educators' attention that, at the same time as attention has been focused on the 'more-than-human' aspects of urban natures, urban space has been the subject of intense restructuring as capital seeks to resolve its on-going 'crisis of accumulation'. Contemporary urbanism is the site of important struggles over meaning, ranging from the resurgence of a popular 'urbanology' which claims to have discovered fundamental laws about people's relationships with cities (Gleeson 2014), an academic 'boosterism' that seeks out and disseminates examples of ideal typical cities that are variously labelled as 'smart', 'liveable', or 'sustainable' (Knox 2014), and more critical accounts of urban processes that explain the urbanization process the final stage of capital's dominance of everyday life (Merrifield 2014). This context requires, I suggest, a wide-ranging discussion of what we as educators should be teaching the next generations about contemporary urban natures.

This then, leads to a final point about the way in which the 'deconstructive' urge that is found in the literature cited here, which seeks to challenge the unity and fixity of categories such as 'urban', 'childhood' and 'nature' relates to more orthodox and institutional notions of the 'schooled child'. In general, the effects of the theoretical movements discussed in this paper seek to interrupt and challenge the assumptions about childhood upon which schooling is based (Diaz's paper in this collection on the complex relations of children completing correspondence courses in rural parts of British Columbia is a good example). The practical problem then faced by environmental educators is then, what is to be done? How do we proceed in 'modern' schools and educational settings when the very basis of what it means to be a 'child' has been undermined? It is surely telling then, that many of the pedagogical interventions discussed in the emerging literature around 'urban/childhood/nature' have been in the margins of the formal curriculum (e.g. one off projects) or in informal settings. If society is embarked on a project of 'green transformation', it is unlikely that schools as institutions will disappear any time soon, and the challenge is how the radical insights of the literature discussed in this paper can be made to inform mainstream educational practice.

Disclosure statement

No potential conflict of interest was reported by the author.

References

Aries, P. 1967. *Centuries of Childhood*. Harmondsworth: Penguin.
Benton-Short, L., and J. R. Short. 2009. *Cities and Nature*. London: Routledge.
Braun, B. 2015. "New Materialisms and Neoliberal Natures." *Antipode: A Journal of Radical Geography* 47 (1): 1–14.
Buckingham, D. 2000. *After the Death of Childhood?* Cambridge: Polity.
Burman, E. 1994. *Deconstructing Developmental Psychology*. London: Routledge.
Castree, N. 2005. *Nature*. London: Routledge.
Clark, N. 2011. *Inhuman Nature: Sociable Life on a Dynamic Planet*. London: Sage.
Cronon, W. 1991. *Nature's Metropolis. Chicago and the Great West*. New York, NY: W.W. Norton.
Davis, M. 2003. *Ecology of Fear. Los Angeles and the Imagination of Disaster*. London: Picador.
Franklin, A. 2017. "The More-than-Human-City." *Sociological Review* 65 (2): 202–17.
Gandy, M. 2002. *Concrete and Clay: Re-Working Nature in New York City*. Cambridge, MA: MIT Press.
Gleeson, B. 2014. *The Urban Condition*. London: Routledge.
Harvey, D. 1996. *Justice, Nature and the Geographies of Difference*. Oxford: Blackwell.

Hendrick, H. 2003. *Child Welfare: Historical Dimensions, Contemporary Debate*. Bristol: Policy Press.

Henriques, J., W. Hollway, C. Unwin, C. Venn, and V. Walkerdine. 1984. *Changing the Subject: Psychology, Social Regulation and Subjectivity*. London: Routledge.

Heynen, N. 2009. "Bending the Bars of Empire from Every Ghetto for Survival: The Black Panther Party's Radical Antihunger Politics of Social Reproduction and Scale." *Annals of the Association of American Geographers* 99 (2): 406–422.

Heynen, N., M. Kaika, and E. Swyngedouw, eds. 2006. *In the Nature of Cities*. London: Routledge.

Heynen, N. 2014. "Urban Political Ecology I: The Urban Century." *Progress in Human Geography* 38 (4): 598–604.

Hinchliffe, S. 2007. *Geographies of Nature*. London: Sage.

Hinchliffe, S., M. Kearnes, M. Degen, and S. Whatmore. 2005. "Urban Wild Things: A Cosmopolitical Experiment." *Environment and Planning D: Society and Space* 23: 643–658.

James, A., C. Jenks, and A. Prout. 1998. *Theorizing Childhood*. Cambridge: Polity.

Kaika, M. 2005. *City of Flows: Modernity, Nature and the City*. London: Routledge.

Klingle, M. 2007. *Emerald City: An Environmental History of Seattle*. New Haven, CT: Yale University Press.

Knox, P. 2014. *Atlas of Cities*. Princeton, NJ: Princeton University Press.

Latham, A., D. McCormack, K. McNamara, and D. McNeill. 2009. *Key Concepts in Urban Geography*. London: Sage.

Lloro-Bidart, T. 2015. "A Political Ecology of Education in/for the Anthropocene." *Environment and Society: Advances in Research* 6: 128–148.

Marston, S., J.-P. Jones III, and K. Woodward. 2005. "Human Geography without Scale." *Transactions of the Institute of British Geographers* 30 (4): 416–432.

Merrifield, A. 2014. *The New Urban Order*. London: Pluto Press.

Messenger-Davies, M. 2010. *Children, Media and Culture*. Maidenhead: Open University Press.

Moss, P., G. Dahlberg, L. Olsson, and M. Vanderbroek. 2016. *Why Contest Early Childhood?* Accessed September 30, 2016. https://www.routledge.com/education/posts/10150?utm_source=shared_link&utm_medium=post&utm_cam

Pacini-Ketchabaw, V. 2013. "Frictions in Forest Pedagogies: Common Worlds in Settler Colonial Spaces." *Global Studies of Childhood* 3 (4): 355–365.

Pacini-Ketchabaw, V., and A. Taylor, eds. 2015. *Unsettling the Colonial Places and Spaces of Early Childhood Education*. Abingdon: Routledge.

Park, R. E., and E. W. Burgess, eds. 1925/1967. *The City*. Chicago, IL: University of Chicago Press.

Pile, S., and N. Thrift, eds. 2000. *City a-Z*. London: Routledge.

Prout, A. 2005. *The Future of Childhood*. London: Sage.

Robbins, P. 2007. *Lawn People: How Grasses, Weeds and Chemicals Make Us Who We Are*. Philadelphia, PA: Temple University Press.

Shillington, L. 2015. *Birds Are for Girls? What Children's Media Teaches Kids about Nature and Cities*. Accessed August 11, 2016. http://www.thenatureofcities.com/2015/05/10

Shillington, L., and A. M. Murnaghan. 2016. "Urban Political Ecologies and Children's Geographies: Queering Urban Ecologies of Childhood." *International Journal of Urban and Regional Research*. doi:10.1111/1468-2427.12339.

Short, J. R. 2006. *Urban Theory: A Critical Assessment*. Basingstoke: Palgrave Macmillan.

Smith, N. 1984. *Uneven Development: Nature, Capital and the Production of Space*. Oxford: Blackwell.

Smith, N. 1993. "Homeless/Global: Scaling Places." In *Mapping the Futures: Local Cultures, Global Change*, edited by J. Bird, B. Curtis, T. Putnman, G. Robertson and L. Tickner, 87–119. London: Routledge.

Smith, S. 1994. "Urban Geography in a Changing World." In *Human Geography: Society, Space, and Social Science*, edited by D. Gregory, R. Martin and G. Smith, 231–251. Basingstoke: Macmillan.

Somerville, M. 2013. "The Natures/Cultures of Children's Place Learning Maps." *Global Studies of Childhood* 3 (4): 407–417.

Somerville, M. 2014. "Entangled Objects in the Cultural Politics of Childhood and Nation." *Global Studies of Childhood* 4 (3): 183–194.

Swyngedouw, E. 1996. "The City as a Hybrid: On Nature, Society and Cyborg Urbanization." *Capitalism Nature Socialism* 7 (2): 65–80.

Taylor, P. J. 1981. "Geographical Scales within the World-Economy Approach." *Review* 5: 3–11.

Taylor, A. 2013a. *Reconfiguring the Natures of Childhood*. London: Routledge.

Taylor, A. 2013b. "Caterpillar Childhoods: Engaging the Otherwise Worlds of Central Australian Aboriginal Children." *Global Studies of Childhood* 3 (4): 366–379.

Taylor, A. 2014. "Settler Children, Kangaroos and the Cultural Politics of Australian National Belonging." *Global Studies of Childhood* 4 (3): 169–182.

Ward, C. 1978. "The Child in the City." *Society* 15 (4): 84–91.

Whatmore, S. 2002. *Hybrid Geographies*. London: Sage.

Wolch, J. 1996. "Zoöpolis." *Capitalism Nature Socialism* 7 (2): 21–47.

Wolch, J. 2008. "Green Urban Worlds." *Annals of the Association of American Geographers* 97 (2): 373–384.

Wolch, J., and J. Emel, eds. 1998. *Animal Geographies: Place, Politics and Identity in the Nature-Culture Borderlands*. London: Verso.

Wyness, M. 2006. *Childhood and Society: An Introduction to the Sociology of Childhood*. Basingstoke: Palgrave Macmillan.

Index

Note: *Italic* page numbers refer to figures and page numbers followed by "n" denote endnotes.

For Product Safety Concerns and Information please contact our EU
representative GPSR@taylorandfrancis.com
Taylor & Francis Verlag GmbH, Kaufingerstraße 24, 80331 München, Germany